8TH GRADE SCIENCE

Unit 9

Ecosystem Changes

AF876483

Table of Contents

CHARACTER SPOTLIGHT

OBSERVANT

Giving careful attention to events, conditions, and people around me with perceptive understanding

OBJECTIVES

- Describe and diagram the parts of a watershed.
- Describe the importance of watersheds to humans.
- Describe aquifers and their role in making and storing clean groundwater.
- Understand how humans can impact watersheds, aquifers, and the freshwater supply.
- Describe water scarcity.
- Compare and contrast point and nonpoint pollution sources.
- Describe the interactions between biotic and abiotic factors in different freshwater ecosystems.
- Infer how short-term changes can affect populations of different freshwater organisms.
- Understand how humans depend on the ocean.
- Describe the different open ocean ecosystems.
- Understand and describe how humans impact the ocean with garbage, oil spills, and carbon dioxide.
- List solutions for cleaning up oil pollution.
- Identify key challenges that exist when trying to predict and manage change within an ecosystem.
- Understand the role that scale, proportion, and quantity play in ecosystem structure and performance.
- Explain the role of engineering in addressing and reducing the impacts of ecosystem changes.
- Understand how engineers help protect and restore ecosystems.
- Explain the role of an environmental engineer.
- Describe the contributions of Dr. Ellen Swallow Richards.
- Identify the impact of Dr. Ellen Swallow Richards in the field of environmental engineering.

Character Spotlight

It's the little details that are vital. Little things make big things happen.

—John Wooden

SECTION ONE

Objectives:

- Describe and diagram the parts of a watershed.
- Describe the importance of watersheds to humans.
- Describe aquifers and their role in making and storing clean groundwater.
- Understand how humans can impact watersheds, aquifers, and the freshwater supply.
- Describe water scarcity.
- Compare and contrast point and nonpoint pollution sources.

Vocabulary:

aquifer *[AK-wuh-fer]* – an underground rock formation that is able to hold groundwater

confluence *[KON-floo-uhns]* – the place where two or more streams or rivers begin to flow together

groundwater – water that collects in the cracks of soil and rock under the ground

headwaters – the beginning of a river

mouth – the place where a stream or river meets an ocean

percolate *[PUR-kuh-leyt]* – to flow slowly through a porous material

permeability *[pur-mee-uh-BIL-i-tee]* – a measure of how well water can move through the pores, or holes, in a substance

porosity *[paw-ROS-i-tee]* – a measure of how many holes are in a substance

recharge – the process of adding water to aquifers through precipitation that percolates into the ground

seep – to slowly flow downward

surface water – water that comes from precipitation and collects in rivers, lakes, ponds, and other bodies of water

water scarcity – an imbalance that occurs when the demand for clean, fresh water is greater than the supply in an area

water table – an underground area where the rocks, gravel, sand, and clay are completely filled with water

watershed – a region of land where all the water is drained into the same water body

1. WATERSHEDS

In the last Unit, we learned about land ecosystems. In this Unit, we will look at aquatic (water) ecosystems. Aquatic ecosystems are divided into two broad categories—fresh water and marine (salt water).

There are two types of fresh water—groundwater and surface water. **Groundwater**, like the name implies, is water that collects in the cracks of soil and rock under the ground. **Surface water** is water that comes from precipitation (rain, snow, sleet) and collects in rivers, lakes, ponds, and other water bodies.

All organisms need water to survive. Many organisms, including humans, need fresh water. Clean, fresh water is one of the earth's greatest natural resources.

The supply of usable fresh water is limited. The earth's surface is 70% to 80% water, and 97% of that water is salt water. Look at **Fig. 1.1**. This picture by the United States Geological Survey (USGS) shows all the world's water sources combined in a sphere. Do not be fooled by the apparent small size of the water. The largest ball of water, representing the world's oceans and that is shown on top of the United States, would also reach 860 miles into the atmosphere.

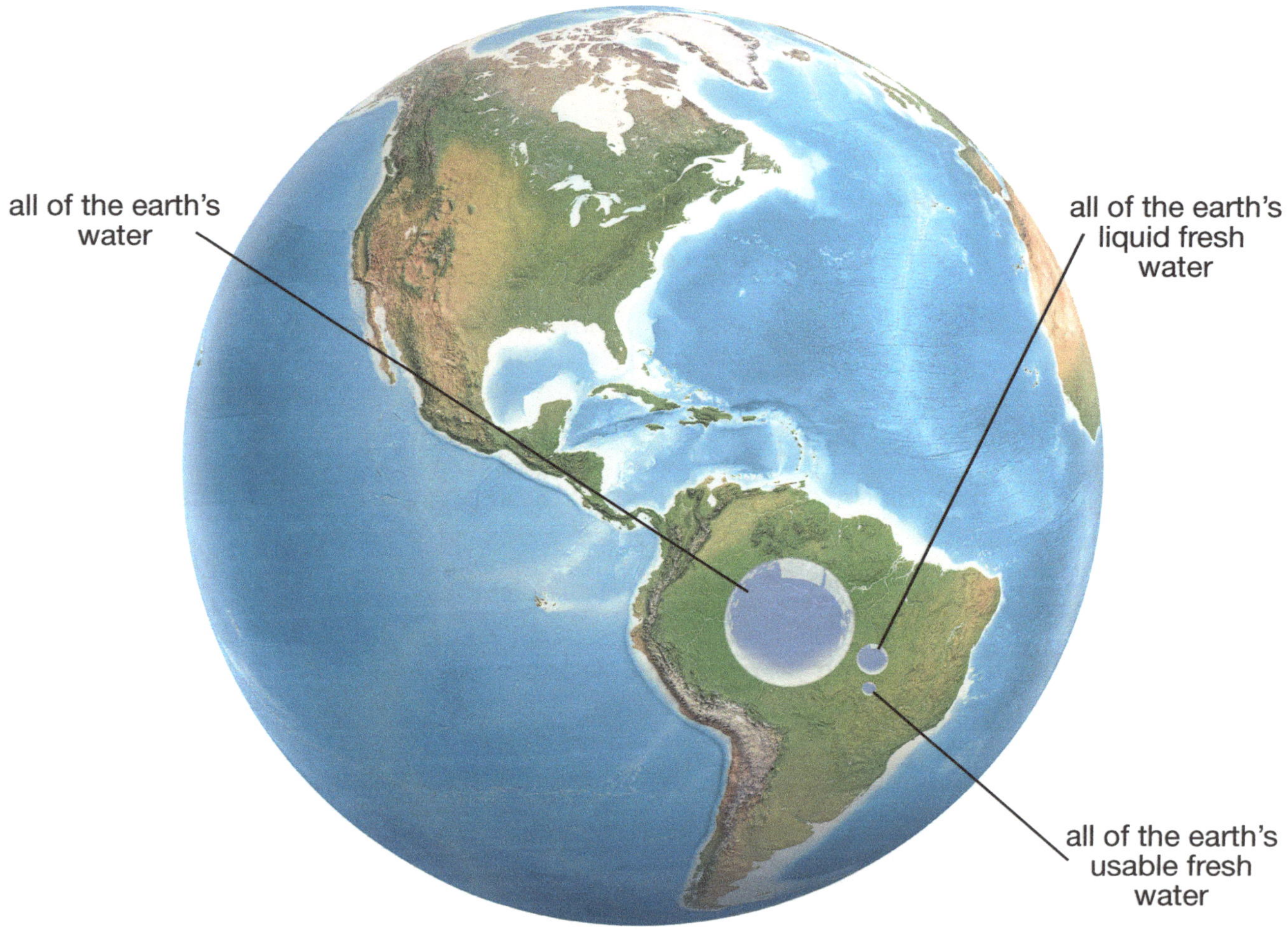

Fig. 1.1 Amount of water on the earth

Fig. 1.2 The main watersheds of North America

Source: Mountain High Maps® Copyright © 1993-2020 Digital Wisdom®, Inc.

The next smaller bubble represents the earth's liquid fresh water. (Remember, 70% of fresh water is in solid ice form within glaciers and ice sheets.) The liquid freshwater bubble includes all groundwater, swamp water, rivers, and lakes. This sphere would reach 170 miles into the atmosphere. Most of the water in this sphere is too deep inside the ground to be used by humans, so the tiniest sphere was created to show surface fresh water that is usable.

The smallest bubble shows how much fresh water is easily accessed by humans in streams and lakes. It reaches 35 miles into the atmosphere. This seems very small compared to the other bubbles. However, if you were to come across a ball of water that had a diameter of 35 miles, that sphere would reach as high as almost six Mount Everests stacked one on top of another.

WATERSHED BASICS

Before studying individual freshwater ecosystems, it is important to first look at how the freshwater ecosystems interact together. The world's liquid fresh water is contained in watersheds. **Watersheds** are regions of land where all the water eventually reaches the same water body. Watersheds describe how water flows through the land, **seeps** (slowly flows downward) into the ground, collects in lakes, streams, larger rivers, and eventually meets the oceans. Because the water is drained, watersheds are also often referred to as drainage basins.

The shape of watersheds is determined by topography, or the land features of the region (**Fig. 1.2**). Water travels downhill, and the shapes and paths of the streams are determined by the sides of the watershed. These sides include ridges, hills, or mountains of the area. These watersheds can be very large or very small and are generally bounded by mountains or hills. One watershed could be made up of thousands of square miles of land in which the water drains into the ocean or only a few square miles in size in which the water drains into a pond. In fact, within one main watershed there can be hundreds of smaller watersheds.

While watersheds consist of a relationship between water and land, it is important to know that more is involved than just water and land. Watersheds also contain people and the stuff people bring to the water. If not properly contained, anything that ends up anywhere in a watershed can end up in streams, rivers, lakes, and oceans. This includes oil leaking out of cars and trash. Just like ecosystems, damage to any one part of the watershed can be damaging to the whole system.

To understand how a watershed works, think of it as an upside-down umbrella (**Fig. 1.3**). The sides of the umbrellas are like the ridges, and the spokes are like the other topographical features of an area, such as slopes and depressions. All the water that falls into the umbrella will eventually flow to the lowest point at the bottom of the umbrella. If the water does not flow or fall inside the watershed or umbrella, then it is not part of that watershed. A large umbrella is a great example of a main watershed. Now imagine that there are numerous smaller upside-down umbrellas stacked on top of the large one, each containing a hole on the bottom. Water from these smaller umbrellas drains into the large umbrella through the holes on the bottom. This is similar to smaller watersheds draining into larger ones.

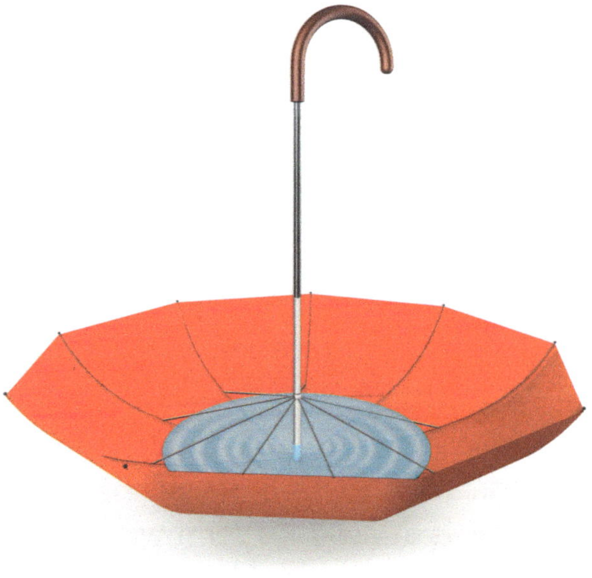

Fig. 1.3 A watershed is like an upside-down umbrella. Everything that falls into the umbrella eventually flows and collects at the lowest point.

Water at the beginning of a watershed usually flows from the highest point of topography (**Fig. 1.4**). The highest point is usually the beginning of a river, which is also known as the headwaters. These **headwaters** are usually the small streams of the river. The place where two or more streams or rivers begin to flow together is called **confluence**, and the place where the streams and rivers meet an ocean is called the **mouth**.

Although this is the typical path of water, not all water in a watershed simply makes its way into a stream and eventually to the mouth of an ocean. Some water is absorbed and transpired by plants. Transpiration is somewhat similar to animals sweating. Plants transpire water to keep cool. Some water seeps through the ground to be stored. Other water gets lost to evaporation from the Sun, and some gets used by humans for drinking and agricultural purposes.

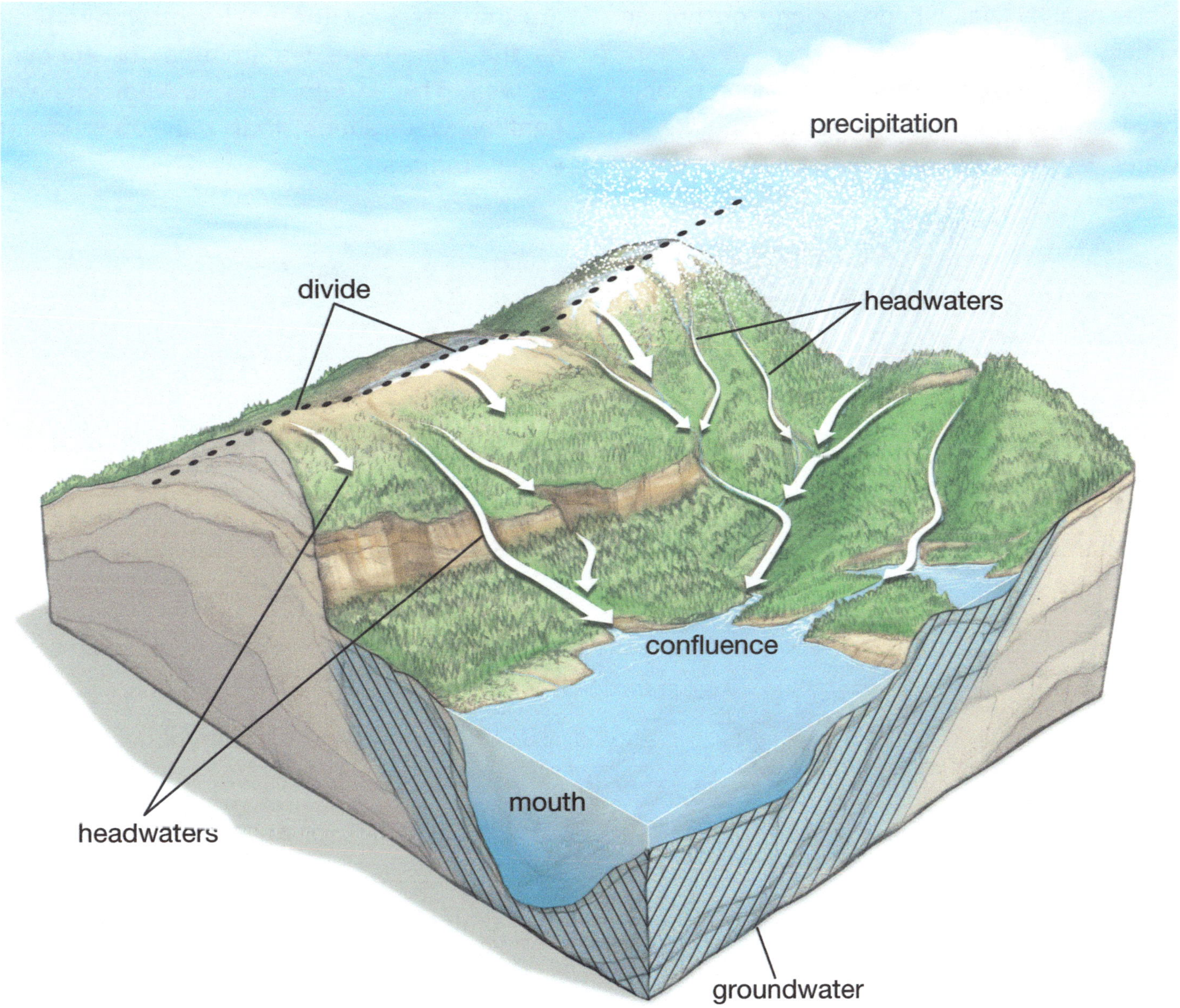

Fig. 1.4 Watershed

Human Interdependence and Impact on Watersheds

Humans are not the only ones who rely on healthy watersheds. Organisms in terrestrial ecosystems need adequate and safe water. Aquatic ecosystems need healthy water so that all the organisms living in the water can thrive. This type of relationship between humans and watersheds is one of interdependence. Interdependence means two or more people or things depend or rely on each other. Organisms in all ecosystems need a healthy and adequate water supply, while healthy watersheds depend on human activity to remain healthy.

We will discuss different impacts people have on groundwater and surface bodies of water in the next several Lessons. People use water for drinking, cooking, bathing, agriculture, and business. Other organisms also depend on healthy water. If a small watershed contains contaminated water, eventually the pollution will reach the larger watershed and the ocean. In this way, even the smallest watersheds can affect the entire freshwater system.

The future of the earth's fresh water lies largely with humans and how we use or abuse this natural resource. Watersheds hold the fresh water humans need to survive and thrive. Plants and animals are impacted by the health of watersheds as they need the watersheds for survival as well. This is why it is so vital to have healthy watersheds that can offer clean water sources.

REVIEW

Match the terms with the definitions.

1.1) _____ confluence

1.2) _____ groundwater

1.3) _____ headwaters

1.4) _____ mouth

1.5) _____ seep

1.6) _____ surface water

1.7) _____ watershed

A. to slowly flow downward

B. water that comes from precipitation and collects in rivers, lakes, ponds, and other water bodies

C. the place where two or more streams or rivers begin to flow together

D. the place where a stream or river meets an ocean

E. the beginning of a river

F. a region of land where all the water is drained into the same body of water

G. water that collects in the cracks of soil and rock under the ground

Write the correct terms on the diagram labels.

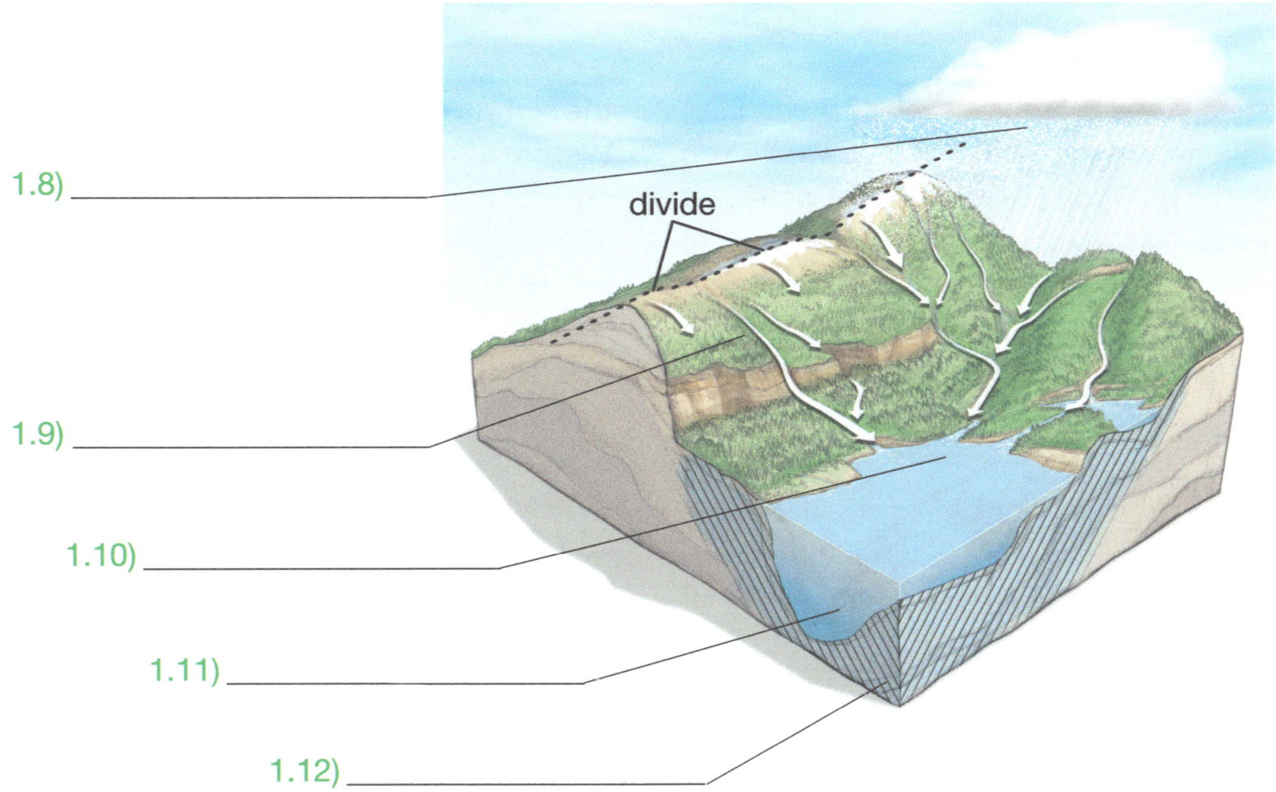

Write T for True or F for False. If False, rewrite the statement to make it True.

1.13) _____ Watersheds hold salt water.

____

____

1.14) _____ What happens in a small watershed has no impact on the larger watershed to which it connects.

____

____

1.15) _____ Humans are interdependent on watersheds.

____

____

1.16) _____ If not properly contained, anything that ends up anywhere in a watershed can end up in streams, rivers, lakes, and oceans.

____

____

CHECK ☐ CORRECT ☐ RECHECK ☐

2. AQUIFERS

Imagine you are at the beach and you start digging a hole in the sand. How far would you have to dig until you found a pool of water? At the beach, you would most likely not have to dig very far. However, the concept holds true for lands farther inside a watershed system. If it is the right kind of land and rocks, and if you dig deep enough, you will find the rocks, gravel, sand, and clay completely filled with water. This water-soaked place is called the **water table** and is part of an **aquifer**. An aquifer is an underground rock formation that is able to hold groundwater (**Fig. 2.1**).

The Importance of Aquifers

Not all water ends up in a pond, lake, stream, or ocean. Some water seeps into the ground and becomes groundwater. Within a watershed, groundwater may be stored in an aquifer. Most of the world's drinking water comes from groundwater located in aquifers.

Once you reach the water table of an aquifer, all the empty spaces between the rocks are filled with water. However, not every rock is created equal. Each rock has a different porosity and permeability. **Porosity** is a measure of how many holes are in a substance. Sand, for example, has greater porosity than clay. **Permeability** is a measure of how well water can move through the pores, or holes, in a substance. This means that each rock within the aquifer has a different number of holes and the ease with

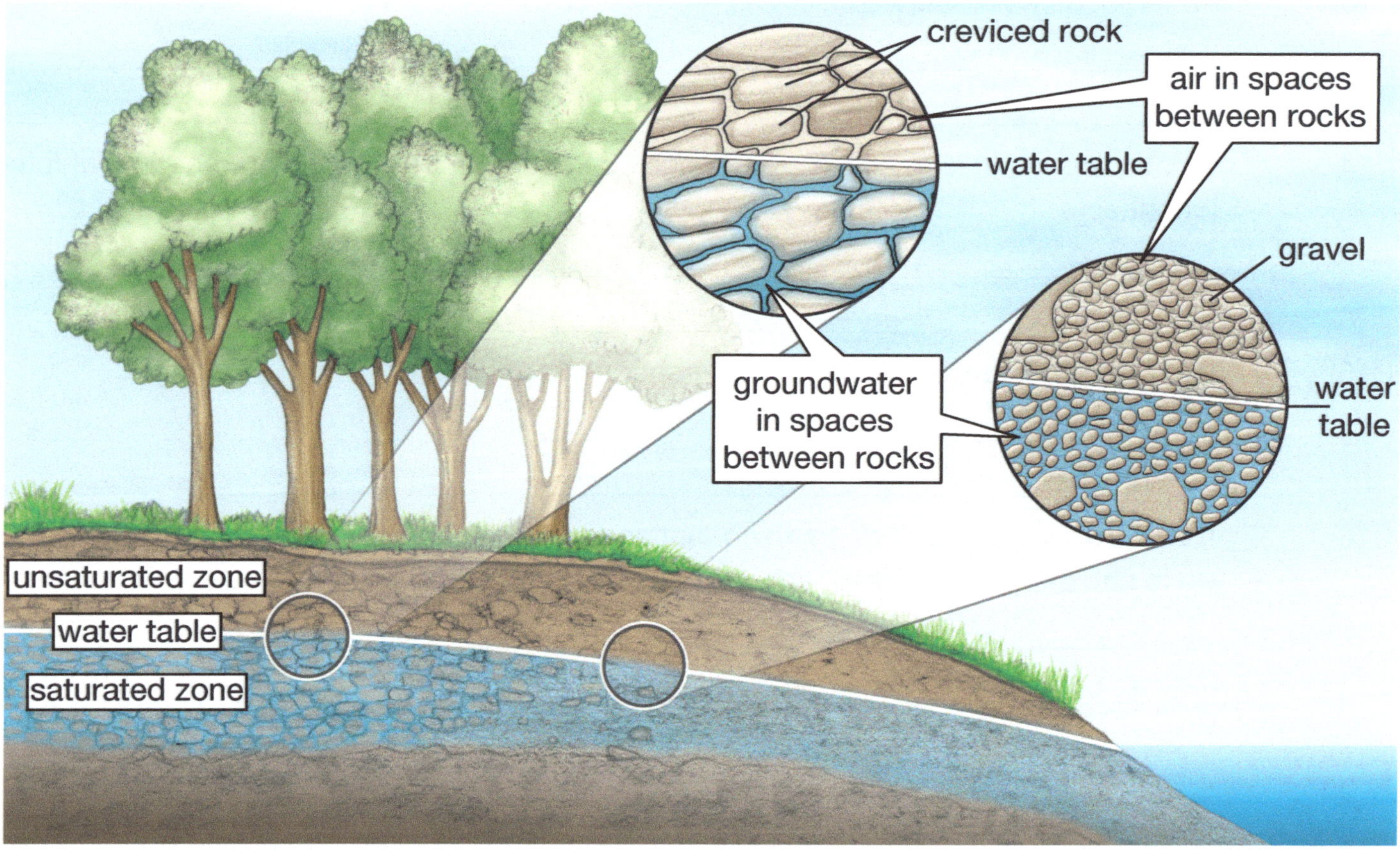

Fig. 2.1 Aquifer zones

which water can move or percolate through these holes is different. To **percolate** is to flow slowly through a porous material.

The rocks that fill aquifers easily allow water to move into wells and springs. A well is a hole drilled into the ground so that fresh water can be pumped out of it. Springs are natural areas where groundwater flows out of the ground. If water is removed from an aquifer by wells and springs, it can also be added to an aquifer.

Water is generally added to aquifers through precipitation that percolates into the ground. This process is called **recharge** and can be slow. How fast an aquifer recharges depends on the type of aquifer, as well as the types of rocks present. Because recharge is slow, if too much water is pumped out of the aquifer through a well, the well can run dry. A neighbor's well also can run dry because the well most likely gets water from the same aquifer. In this way, if someone overpumps the well before the aquifer has a chance to recharge, that person is taking the neighbor's supply of water.

While water in the saturated zone of an aquifer completely fills the space between the rocks, this does not hold true for the unsaturated zone of an aquifer. The unsaturated zone is the area of an aquifer that contains both air and water between the spaces of the rocks. These rocks are above the water table. While water can still be found, which is the reason why plants and trees can grow successfully, it does not take up every available space in the unsaturated zone.

Types of Aquifers

There are two main types of aquifers: unconfined and confined (**Fig. 2.2**).

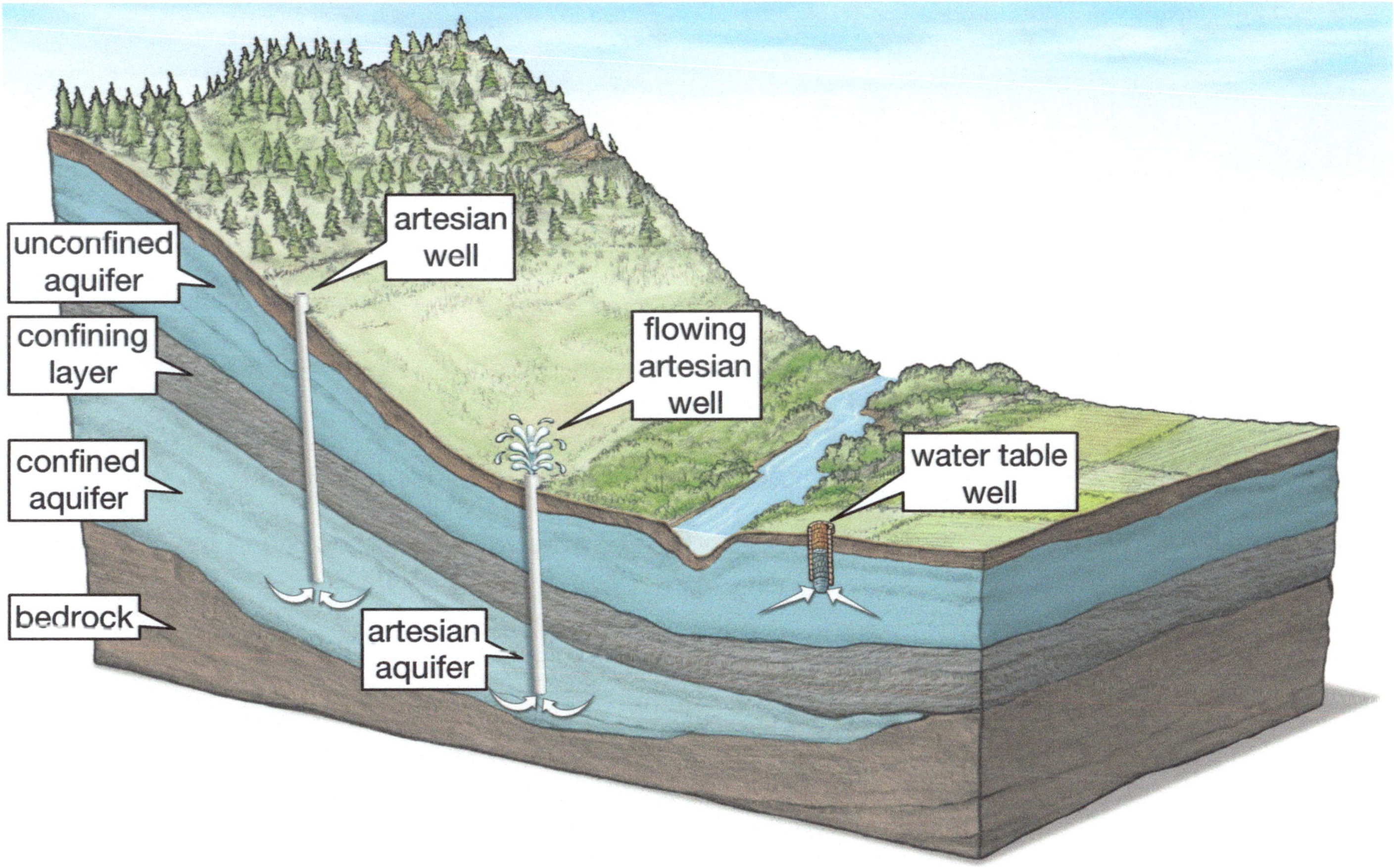

Fig. 2.2 Types of aquifers

An unconfined aquifer is an aquifer in which water seeps into the ground directly above the aquifer. Unconfined aquifers have a faster recharge rate because there is more surface area for the water to enter the aquifer. Unconfined aquifers are also in greater danger of becoming polluted because polluted water can easily enter these aquifers at many different points.

A confined aquifer is an aquifer in which a surface with no holes keeps water from seeping into it. This forces the water to enter the aquifer at a point farther away. One special type of confined aquifer is called an artesian aquifer. An artesian aquifer is shaped so that the water ends up under a great deal of pressure. If a well is dug into an artesian aquifer, water will burst forth from the well as it is released from all the pressure. Because of the great release of pressure, wells dug into artesian aquifers do not need pumps, and the water coming up often puts on a beautiful, dramatic show depending on just how much pressure the water was under.

As you can see, aquifers are incredibly important to the watershed system. Aquifers hold groundwater, one of the greatest sources of drinkable water, and the health of groundwater and aquifers depends upon the health of everything above the ground's surface. As we continue our discussion of freshwater ecosystems, keep the watershed system in mind and how each separate freshwater ecosystem is a vital piece of a watershed.

REVIEW

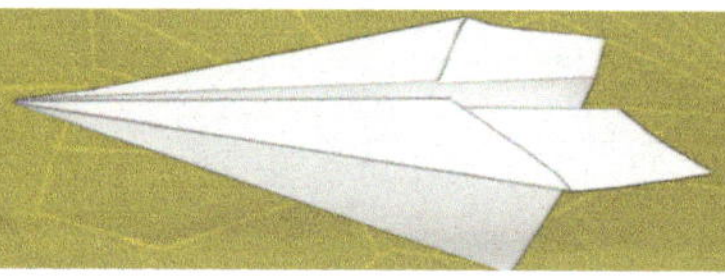

Match the terms with the definitions.

2.1) _____ aquifer

2.2) _____ percolate

2.3) _____ porosity

2.4) _____ permeability

2.5) _____ recharge

2.6) _____ water table

A. to flow slowly through a porous material

B. an underground area where the rocks, gravel, sand, and clay are completely filled with water

C. a measure of how well water can move through the pores, or holes, in a substance

D. the process of adding water to aquifers through precipitation that percolates into the ground

E. an underground rock formation that is able to hold groundwater

F. a measure of how many holes are in a substance

Choose the best answers.

2.7) _____ Groundwater is stored in ___.
A. aquifers
B. lakes
C. rivers
D. oceans

2.8) _____ Groundwater can be accessed through ___.
A. ponds
B. wells
C. lakes
D. streams

2.9) _____ If an aquifer has very porous rocks, will the well water move faster or slower?
A. faster
B. slower
C. the water flows with the use of pumps

2.10) _____ Why does an artesian aquifer not need a well with a water pump?
A. the water is under a great deal of pressure
B. the water is under more gravity
C. the water flows naturally to a river
D. the water flows with the use of other pumps

2.11) _____ Why is groundwater important?
A. used as drinking water
B. supplies water to streams
C. used for transportation
D. breaks down rocks

Write U for Unconfined or C for Confined next to the descriptions of aquifers.

2.12) _____ has a faster recharge rate

2.13) _____ is more easily polluted

2.14) _____ an aquifer in which a surface with no holes keeps water from seeping into it

2.15) _____ an aquifer in which water seeps into the ground directly above the aquifer

2.16) _____ an aquifer in which water must enter at a point farther away

CHECK ☐ CORRECT ☐ RECHECK ☐

3. WATER SCARCITY AND WATERSHED POLLUTION

Clean, healthy water comes from watersheds. Two major ways that humans impact watersheds is the amount of water used and pollution. Using too much water in a watershed can lead to water scarcity. **Water scarcity** is an all-encompassing term that refers to an imbalance that occurs when the demand for clean, fresh water is greater than the supply in an area. Water scarcity deals with water use at the local level. Each area where people and animals live has some type of water source. If people use too much of their local resource or pollute it, water scarcity results. There are three components to water scarcity:

1 – Lack of water

Either an area is experiencing a drought, the area has never had a good water supply, or water from rivers and aquifers is getting used up faster than it can be recharged for homes, businesses, and agriculture.

2 – Lack of tools to access available fresh water

Some areas have underground water but lack wells to access it. In other areas there may be a river, but communities cannot afford pipes to bring the water to the residents. Some people in Africa spend five to six hours per day walking to retrieve water for their households.

3 – Polluted and unsafe water

Once water becomes polluted, water scarcity results, and people then lack clean water for drinking, bathing, and eating. Using polluted water increases the chances of contracting diseases, such as cholera, typhoid fever, dysentery, plague, and trachoma. However, storing water long term can also be unsafe, as stored water is prone to bacteria contamination and can become a breeding ground for malaria-carrying mosquitoes.

Water scarcity most often refers to a shortage of water, especially in areas that do not receive much precipitation. Many of the places that have the least amount of water are experiencing very high population growth, adding strain to an already weak water supply.

Pollution can cause water scarcity.

Each year, the human population increases by about 80 million people, and all these people need water. Competition for water is present at all levels and is projected to increase in the future.

Some regions in the United States are headed toward water scarcity if habits are not changed. Here is a list of America's regions most at risk of suffering from water scarcity. This data is from the Columbia University Water Center.

- Washington, D.C., metropolitan area
- New York City metropolitan area
- California area from San Diego to Santa Barbara and inland

- North and South Dakota agricultural belt
- Nebraska agricultural belt
- Illinois
- Arkansas area of the lower Mississippi belt
- North Texas agricultural belt
- Ohio agricultural regions
- Minnesota agricultural regions

One way to keep water supplies healthy is to keep ecosystems healthy. Ecosystems can filter water and help water be able to seep into the ground. By keeping ecosystems in balance, water can be cleaned and recycled.

POLLUTION

Watersheds and aquifers are also impacted by pollution. It is not safe to drink, swim in, or eat fish out of polluted water. The definition of healthy water according to the Environmental Protection Agency (EPA) is water that is swimmable and fishable.

The runoff in this picture is going down a storm drain. Many storm drains lead to a lake or river.

Sources of Water Pollution

There are two main sources of water pollution: point-source pollution and nonpoint-source pollution. Point-source pollution is pollution that comes from a single source, or point. Point-source pollution is very easy to identify and clean up. Nonpoint-source pollution comes from many different sources and can be difficult to identify and clean up. In the United States, up to 95% of water pollution is from nonpoint sources.

Most nonpoint-source pollution comes from contaminated runoff. Runoff is water that flows over the land's surface. When there is more rain than the ground can hold, the excess water will "run off" onto the roads, grass, and other areas.

To control nonpoint-source pollution, individuals and businesses must adopt practices that will reduce the amount of pollutants entering the runoff into watersheds.

Take a look at different sources of point-source and nonpoint-source pollution (**Fig. 3.1**).

Sources of Point-Source Pollution

- leaking septic tanks
- leaking underground storage tanks that contain gasoline or other similar chemicals
- polluted water from active and abandoned mines
- landfills (where garbage is disposed of) that are not lined
- sewage that is dumped into streams directly by different industries

Sources of Nonpoint-Source Pollution

- chemicals added to roads, such as salt and deicing agents
- oil, gas, and various types of litter on roads or in water
- pesticides, herbicides, and fertilizers that are used in lawns, golf courses, and farms
- feces from farm animals
- rain or snow that contains pollutants from the atmosphere

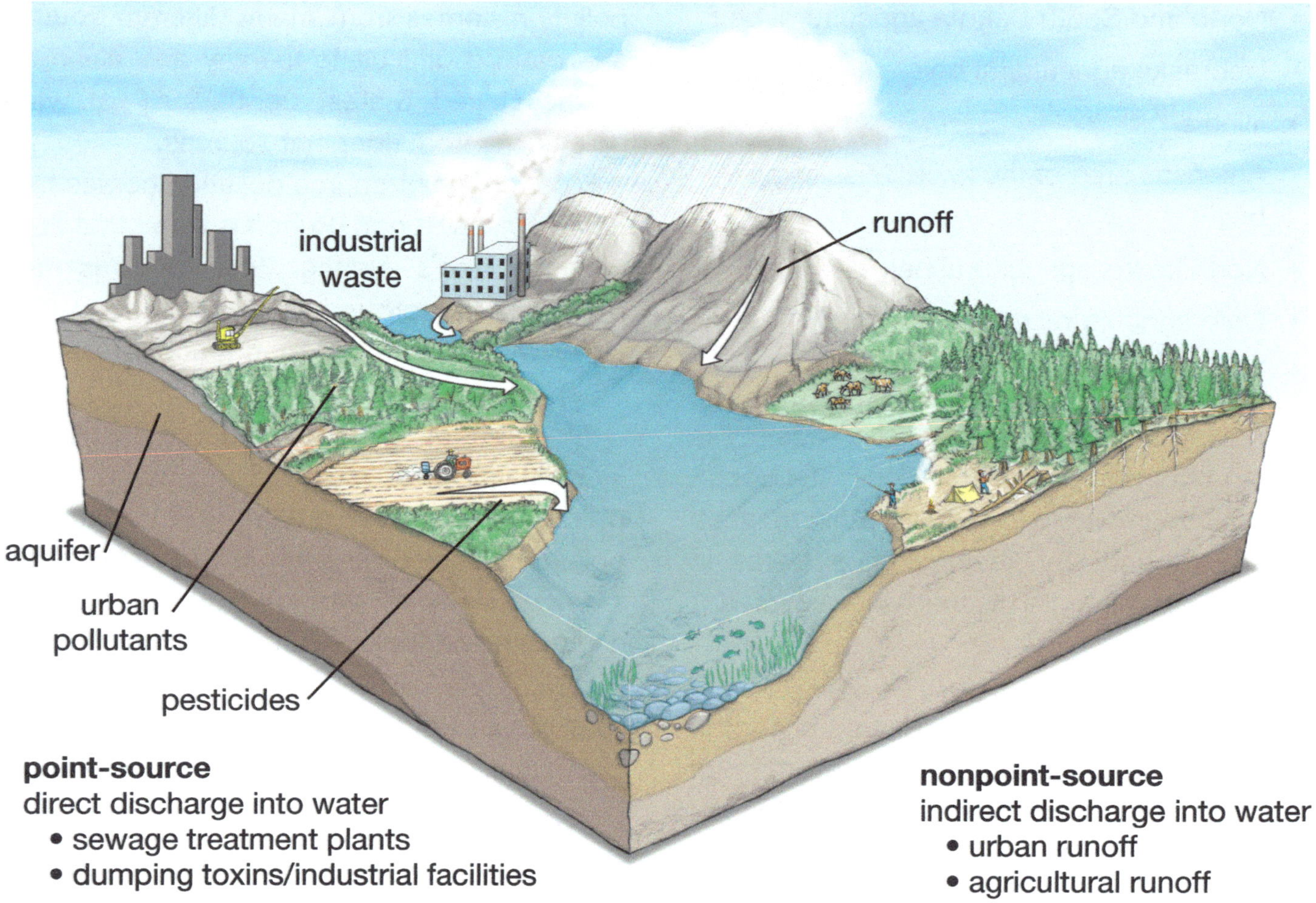

Fig. 3.1 Point-source and nonpoint-source pollution in a watershed

REVIEW

Write N for Nonpoint-source or P for Point-source next to the pollution descriptions.

3.1) _____ feces from farm animals

3.2) _____ a leaking septic tank

3.3) _____ sewage that is dumped into streams directly by different industries

3.4) _____ oil, gas, and various types of litter on roads or water

3.5) _____ pollution that comes from a single source

3.6) _____ pollution that comes from contaminated runoff

Types of Pollutants

There are many different types of pollutants, including chemicals, bacteria, and waste materials. Some of the most common types of pollutants that can get into water are nutrients, sediments, toxic chemicals, heavy metals, and oil and grease. Pollutants can enter surface water via runoff or groundwater percolating. Groundwater pollution is much more difficult to clean up than surface water pollution. If the rocks and gravel in an aquifer become contaminated, they will pollute the clean water that seeps down into the aquifer.

Wastewater

Wastewater is water that contains the waste (from toilets and drains) from homes and industries. Most wastewater from sinks and toilets is biodegradable (can break down). Waste that is not biodegradable becomes sewage sludge, which is solid and highly toxic. Wastewater also comes from industrial sites. Thermal pollution occurs when very hot water used in industrial cooling systems is discharged directly into a lake or river. The hot water can lead to fish kills. A fish kill is when hundreds to thousands of fish die around the same time. Even just warm water can raise the temperature enough to cause oxygen levels to drop and disrupt aquatic ecosystems. Up to 70% of industries in developing countries dump their wastewater directly into waterways.

Algae Blooms

Algae blooms are events where the population numbers of algae in a body of water increase rapidly due to too many nutrients entering the body of water. The increase in nutrients causes an increase in the reproduction of algae.

Algae blooms

When algae die, they decompose. Part of the decomposition process requires oxygen. The decomposing algae can take the oxygen in the water, making the levels too low for fish and other aquatic organisms and causing fish kills.

Fish kills from lack of oxygen due to decomposing algae

Also, as dead and decaying algae matter are added to the aquatic ecosystem, sediment layers build up. Too many algae blooms will eventually cause an aquatic ecosystem, such as a lake, to dry up and become land.

Many algae blooms are caused by human activities. These activities include agriculture, gardening, raising livestock, and sewage. All these activities can put excess nutrients into the environment. The main nutrients that cause algae blooms are nitrogen and phosphorus. Fertilizers and wastewater are the main sources of nitrogen and phosphorus. Fertilizers come from farms, homes, gardens, and golf courses. Phosphorus comes from laundry detergents and dishwashing soaps that enter the groundwater or surface water through runoff.

Disease

Pathogens (organisms that cause disease) enter water when livestock pass through streams, when runoff is polluted, and when feces are dumped directly into a water body. One gram of feces can potentially contain high levels of these dangerous substances:

- 10 million viruses
- 1 million bacteria
- 1,000 parasite cysts
- 100 parasite eggs

Many diseases and deaths can be prevented through drinking safe, clean water that is free of feces. Pollution comes from many sources. We need to identify and stop pollution from point sources while reducing the amount of nonpoint-source pollution entering runoff.

Conservation Efforts

There are many efforts to attempt to limit pollution and reduce water use. Personal reduction of water use can occur with the following actions:

- Take shorter showers.
- Turn off water when not in use.
- Use energy-efficient appliances.
- Water plants and gardens early or late in the day.
- Use special low-flow nozzles and faucets.

Pollution reduction efforts include these steps:

- Follow water conservation activities.
- Do not throw chemicals or medicines down the sink or drain.
- Use environment-friendly household products.
- Do not overuse pesticides or fertilizers.
- Do not litter into rivers, lakes, or oceans.

Washing your car on the lawn instead of the driveway is one way to keep the phosphorus in soap from entering runoff.

Choose the best answers.

3.7) _____ Which type of water is harder to clean up if polluted?

A. groundwater
B. surface water

3.8) _____ Algae blooms are caused by ___.

A. excess nutrients
B. water scarcity
C. excess pesticides
D. water abundance

3.9) _____ When very hot water used in industrial cooling systems is discharged directly into a lake or river, it is called ___.

A. pathogen pollution
B. thermal pollution
C. groundwater pollution
D. aquifer pollution

Order the algae bloom events using the numbers 1–4.

3.10) _____ fish kills

3.11) _____ high rate of algae reproduction

3.12) _____ excess nutrients in water

3.13) _____ decomposition of algae

Choose the best answers.

3.14) Excess nutrients can come into water from which TWO sources?

☐ fertilizers used on the lawn
☐ soap used to wash a car on the street
☐ removal of more water from aquifers than can be recharged
☐ lack of tools to access available fresh water

3.15) Water scarcity is caused by which THREE factors?

☐ polluted and unsafe water
☐ lack of water
☐ lack of human activity
☐ lack of tools to access available fresh water
☐ lack of biodiversity

3.16) Pathogens enter water by which THREE ways?

☐ when livestock pass through streams
☐ through algae blooms
☐ when rain or snow contains pollutants from the atmosphere
☐ when runoff is polluted
☐ when feces are dumped directly into a water body

CHECK ☐ CORRECT ☐ RECHECK ☐

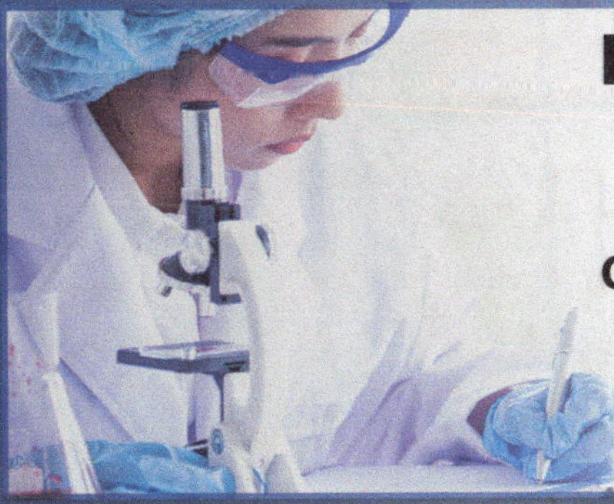

LAB: NUTRIENTS AND ALGAE GROWTH

Complete this lab in the Lab Manual.

Teacher Initial ☐

MY SCORE

QUIZ 1

TEACHER INITIAL

(Each question, 5 points)

Fill in the blanks using words from the box below.

aquifer	water table	watershed	point	nonpoint

1.01) A(n) ______________________________ is a region of land where all the water is drained into the same body of water.

1.02) A(n) ______________________________ is an underground area where the rocks, gravel, sand, and clay are completely filled with water.

1.03) A(n) ______________________________ is an underground rock formation that is able to hold groundwater.

1.04) ______________________________-source pollution is pollution that comes from a single source.

1.05) ______________________________-source pollution is pollution that comes from contaminated runoff.

Choose the best answers.

1.06) Choose TWO ways that excess nutrients can come into water.

☐ fertilizers used on the lawn
☐ soap used to wash a car on the street
☐ rocks with low porosity
☐ removing more water that can be recharged

1.07) _____ ___ is caused by pollution, a lack of water, or lack of tools to access available fresh water.

A. Point-source pollution
B. Runoff
C. Drought
D. Water scarcity

1.08) _____ Aquifers store ___.

A. surface water
B. groundwater
C. runoff
D. sewage

1.09) _____ *Aquifer X has more porous rocks than aquifer Y.*

Which aquifer will have the faster recharge rate?

A. X

B. Y

1.010) _____ *Aquifer M has a faster recharge rate than Aquifer P.*

Which aquifer is more likely to be a confined aquifer?

A. M

B. P

1.011) _____ *Aquifer V is more easily polluted than aquifer W.*

Which aquifer is more likely to be an unconfined aquifer?

A. V

B. W

1.012) _____ *Several industries put their sewage directly into streams.*

What kind of pollution is this?

A. point-source pollution

B. nonpoint-source pollution

C. watershed pollution

D. aquifer pollution

1.013) _____ *Oil and trash enter the storm drain when it rains. The storm drain empties into a river.*

What type of pollution is this?

A. point-source pollution

B. nonpoint-source pollution

C. watershed pollution

D. aquifer pollution

1.014) _____ Is it easier to clean up polluted groundwater or polluted surface water?

A. groundwater

B. surface water

1.015) _____ Which set of events is the correct order for an algae bloom to occur?

A. 1 – Algae reproduce at a high rate.
   2 – Level changes in water cause fish kills.
   3 – Algae decompose and take up oxygen.
   4 – Excess nutrients enter the water.

B. 1 – Excess nutrients enter the water.
   2 – Algae reproduce at a high rate.
   3 – Level changes in water cause fish kills.
   4 – Algae decompose and take up oxygen.

C. 1 – Algae reproduce at a high rate.
   2 – Excess nutrients enter the water.
   3 – Algae decompose and take up oxygen.
   4 – Level changes in water cause fish kills.

D. 1 – Excess nutrients enter the water.
   2 – Algae reproduce at a high rate.
   3 – Algae decompose and take up oxygen.
   4 – Level changes in water cause fish kills.

Match the watershed labels with the letters on the diagram.

1.016) _____ groundwater

1.017) _____ confluence

1.018) _____ precipitation

1.019) _____ mouth

1.020) _____ headwaters

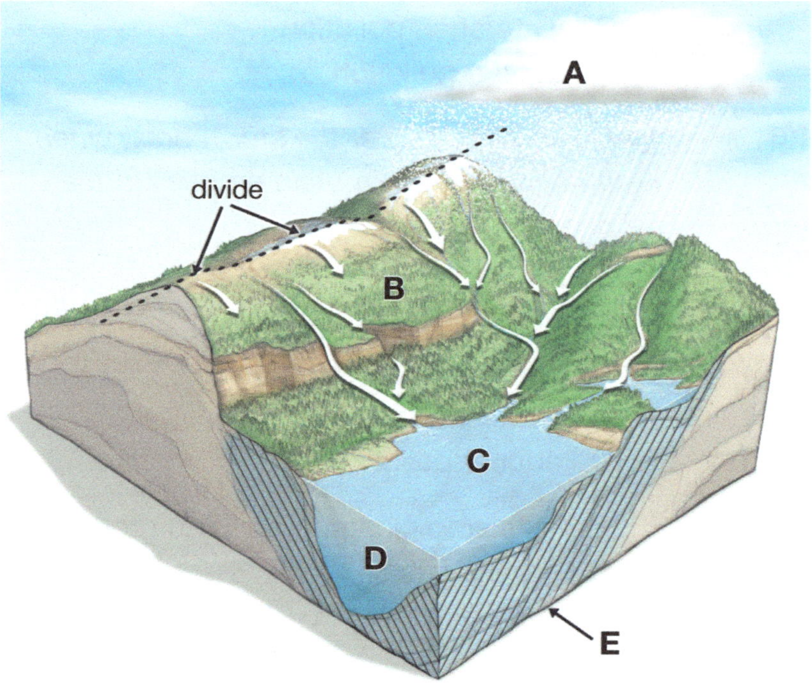

CHECK ____ CORRECT ____ RECHECK ____

SECTION TWO

Objectives:

- Describe the interactions between biotic and abiotic factors in different freshwater ecosystems.
- Infer how short-term changes can affect populations of different freshwater organisms.
- Understand how humans depend on the ocean.
- Describe the different open ocean ecosystems.
- Understand and describe how humans impact the ocean with garbage, oil spills, and carbon dioxide.
- List solutions for cleaning up oil pollution.

Vocabulary:

benthos *[BEN-thahs]* – bottom-dwelling aquatic organisms

biodegradable *[bahy-oh-di-GREY-duh-buhl]* – to be able to be broken down by living organisms into different chemical compounds

biomagnification *[bahy-oh-mag-nuh-fih-KEY-shun]* – how a toxin both moves and increases in concentration through each energy level in a food web

biomass – the total mass of an organism, or the total amount of organisms within an ecosystem

chemosynthesis *[kee-moh-SIN-thuh-sis]* – the process of converting chemicals into an energy form usable by organisms

estuaries – areas where rivers meet the oceans, or where surface fresh water mixes with surface salt water

eutrophic *[yoo-TROF-ic]* **lakes** – lakes that tend to be high in nutrients and low in oxygen

lake – a still body of surface water deep enough that light cannot reach the bottom

nekton – free-swimming aquatic organisms

oligotrophic *[ol-i-goh-TROF-ic]* **lakes** – lakes that lack nutrients but are full of oxygen

plankton – microscopic organisms that make up the bottom of the food chain in aquatic ecosystems

productivity – the rate at which new biomass is produced

riparian *[ri-PAIR-ee-uhn]* **zone** – the area immediately surrounding a lake or river that contains trees, grasses, and other vegetation

river – a stream that contains a large volume of flowing or moving surface water

Character Spotlight

Improve Your Concentration

Learning to concentrate is the key to becoming observant. It allows the mind to connect the things that are being observed to information that already exists within. Therefore, it is important to develop the ability to concentrate on tasks with a razor-sharp focus. How can you improve your powers of concentration?

1. Get an analog clock (one with a sweep second hand). Sit in front of the clock and watch the second hand move around the dial for five minutes. Do not allow yourself to look away from the clock for even a moment. Do this every day for a month.

2. Read a chapter of a book one word at a time. Do not think the thoughts of the book. Instead, read the individual words—and look at every punctuation mark. See if there is a word or punctuation mark that is incorrect.

3. Take a walk in the woods without any electronics. When you get home, write down ten interesting things you observed on your walk.

4. If you find yourself being distracted while trying to do any of these techniques, smile to yourself, get rid of the distraction, and continue the technique, forcing your mind to return to the task at hand.

Almost everything worthwhile that is accomplished is achieved after a period of sustained, uninterrupted thought. Make it a goal to learn to work for long periods of time with sustained attention. If you can do this, you can learn or accomplish anything!

4. FRESHWATER ECOSYSTEMS

Have you ever gone fishing at a lake? Lakes are freshwater ecosystems. Freshwater ecosystems are aquatic environments that contain lower salt levels as compared to marine ecosystems. They include a range of habitats, such as lakes, rivers, and estuaries. Freshwater ecosystems are home to many species of animals and plants, including amphibians, fish, reptiles, and microbes. Freshwater ecosystems provide food and drinking water for humans. They also support a wide variety of industries, from fisheries to agriculture.

WHAT IS A LAKE?

A **lake** is a still body of surface water surrounded by land. The water is deep enough that light cannot reach the bottom. Lakes contain less than 0.01% of the earth's surface water. However, they hold 98% of the earth's fresh liquid water. In comparison, the Antarctic holds 90% of the earth's fresh water, but it is in the form of ice. Lakes are important for life on the earth, and the life found in them helps make up vital ecosystems.

There are two types of lakes: eutrophic and oligotrophic. **Eutrophic lakes** have a lot of nutrients but not much oxygen, while **oligotrophic lakes** have low nutrients but a lot of oxygen. Eutrophic lakes are often found in areas with rich soil and lots of plants and algae, which can lower the amount of oxygen. Oligotrophic lakes are found in colder or

Aquatic Organisms

Aquatic ecosystems, like lakes, are home to diverse groups of living organisms that are classified based on their location and movement within the water (**Fig. 4.1**). **Plankton** are microscopic organisms that drift in the water and include phytoplankton (plant-like protists) and zooplankton (animal-like protists). **Benthos** are bottom-dwelling organisms, such as oysters, starfish, sea worms, and bacteria. **Nekton** are free-swimming organisms, like fish and turtles that can move through the water at will. All of these groups play important roles in the functioning of aquatic ecosystems and interact with each other and the abiotic elements of their environment.

Fig. 4.1 Aquatic organisms are classified by their location and how they move.

higher altitude areas. Eutrophic lakes can be hard for other organisms, like fish, to survive in due to low oxygen levels.

Lake Food Webs and Resources

What are the abiotic and biotic resources that organisms compete for?

Biotic resources include living and once-living organisms. These organisms can be algae, fish, turtles, and aquatic plants. Phytoplankton and aquatic plants are the producers in a lake, while zooplankton, fish, and other organisms are all consumers.

Typical abiotic resources that lake ecosystems need and compete for include:

- temperature range
- oxygen levels
- water levels
- nutrients

Like all organisms in an ecosystem, lake organisms depend on other organisms for food. There can be competition for food sources as well as space (**Fig. 4.2**).

Human Interdependence and Impact on Lakes

People rely on lakes for food and recreation. Organisms both in and out of water rely on lakes for both water and food.

People can impact lakes in the following ways:

- Pollution of lakes can happen when people dump garbage directly into lakes or when wind blows garbage from other areas into the lakes. If a storm drain feeds into a lake, then lakes can be polluted from runoff that has chemicals, excess nutrients, and

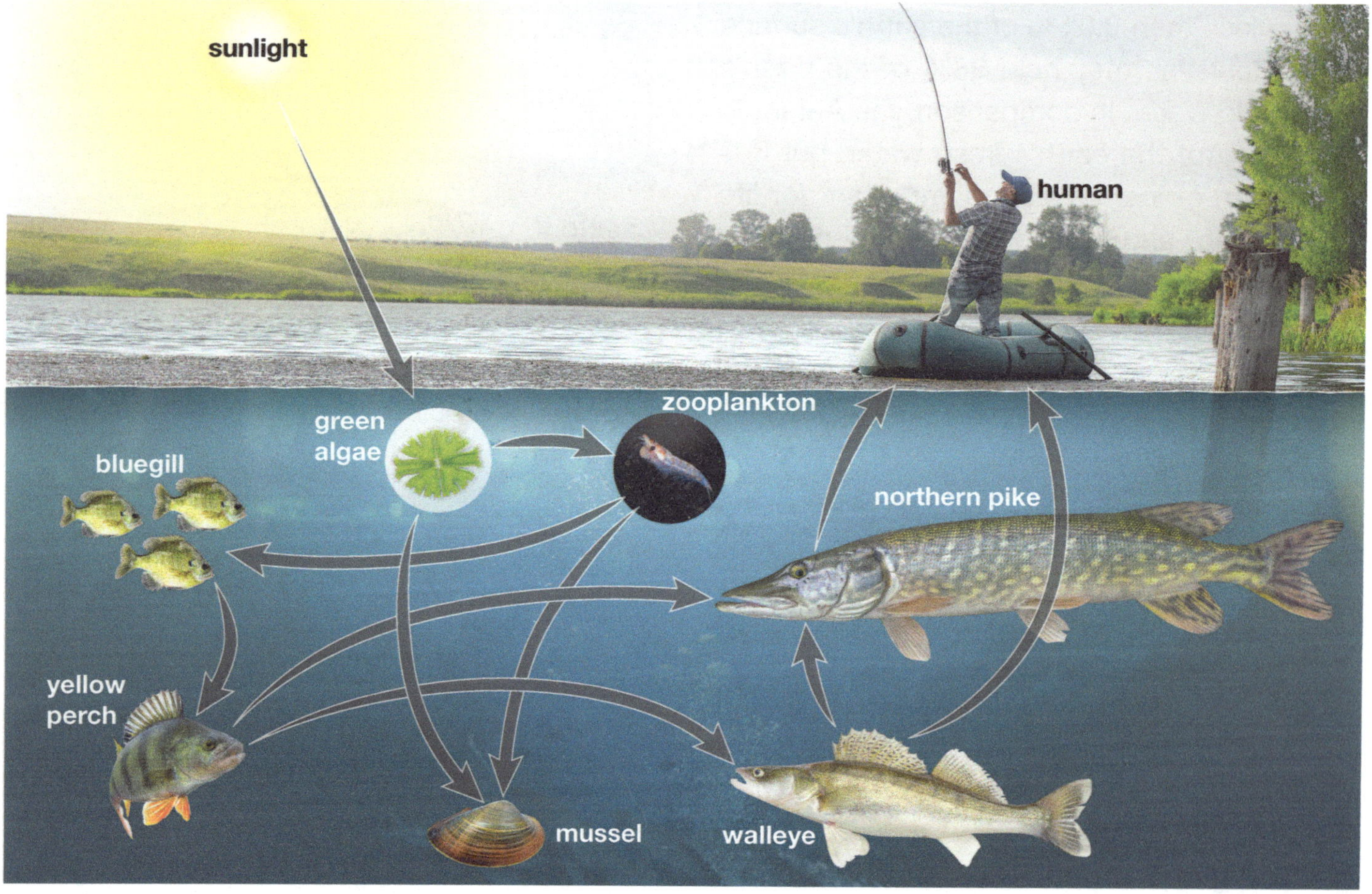

Fig. 4.2 A lake food web

other toxins in it. Pollution often causes population levels of organisms to decrease.

- A drought might naturally cause a lack of water in lakes. However, sediments (small pieces of rock and soil) from erosion can naturally cause a lake to fill up and dry out. People can also withdraw too much water from a lake.
- Excess nutrients can come from runoff or livestock feces. Excess nutrients can cause algae blooms and plankton population levels to increase reducing oxygen and other resources, and causing fish population levels to decrease.

Excess nutrients in lakes can come from cow feces.

- Invasive species can be introduced in several ways. Boats that are not properly cleaned can introduce zebra mussels to different lakes. Some people dump fish from their aquariums into lakes. Both scenarios can cause a new species to take over a lake quickly. In Hawaii, a person dumped a mating pair of piranhas into a lake. The piranha fish quickly took over the lake and killed off the native species of fish. They also made the lake an unpleasant place for people to swim. Invasive species populations usually increase in a new lake environment, reducing resources such as oxygen, while native population levels decrease.

What is a River?

A **river** is a stream that contains a large volume of flowing or moving surface water. The water moves from one place to another, moving dirt, sediments, fish, boats, and people along with it. Rivers also collect the water that is a part of a watershed and carry it to the ocean. Rivers provide a habitat for many different forms of life, including plants, animals, protists, and bacteria.

Rivers flow from high elevations to low ones, making their way through cities, towns, and rural areas before eventually reaching the sea. The water at the beginning of the river is usually clearer and cooler, with higher oxygen levels. However, the water at the mouth of the river is darker and has lower oxygen levels. Rivers shape the land through erosion and can create new land areas called river deltas. **Riparian zones** are areas of vegetation along the banks of a river and are important for maintaining healthy river ecosystems. Trees and plants in riparian zones provide shade and help keep the water cool. If they are removed, the water becomes warmer. Warmer waters cause the amount of dissolved oxygen in rivers to decrease. This is because

A river in the Himalayan Mountains

warmer water cannot hold as much oxygen as colder water. Therefore, if water temperatures increase, there will be less oxygen available for fish and other aquatic creatures to breathe.

Trees and plants in riparian zones also help to stabilize the banks and reduce the risk of erosion. When it rains, the roots help hold the soil in place and keep it from washing into the river. Too much dirt and other particles in the water can make it hard for oxygen to get into the water. This is because the sediments can block sunlight from reaching the bottom of the water, where plants and algae produce oxygen through photosynthesis. When there is less light reaching the bottom, there is less oxygen produced.

Healthy riparian zones are necessary for healthy rivers.

REVIEW

Choose the best answers.

4.1) _____ The area besides the stream that has plants and trees is called the ___.

A. vegetation zone
B. bathyal zone
C. aquatic zone
D. riparian zone

4.2) _____ What might happen to water temperatures and fish populations if the trees and plants are removed?

A. Water temperatures would rise, and oxygen and fish populations would likely increase.
B. Water temperatures would rise, and oxygen and fish populations would likely decrease.
C. Water temperatures would lower, and oxygen and fish populations would likely increase.
D. Water temperatures would lower, and oxygen and fish populations would likely decrease.

4.3) _____ What might happen to the sediment levels and oxygen levels if tree roots are removed from the streams?

A. Sediment levels increase, causing oxygen levels to increase.
B. Sediment levels increase, causing oxygen levels to decrease.
C. Sediment levels decrease, causing oxygen levels to increase.
D. Sediment levels decrease, causing oxygen levels to decrease.

River Food Webs and Resources

Rivers provide abiotic and biotic resources for the populations of organisms living in and nearby the rivers (**Fig. 4.3**). Biotic factors that are important to rivers include the organisms in the water as well as the plants in the riparian zone.

Abiotic factors that organisms need to survive in rivers include:

- pH range
- temperature range
- oxygen levels
- sunlight levels (If the water is especially muddy or full of sediment, light levels will decrease.)

Human Interdependence and Impact on Rivers

People rely on rivers for many reasons:

- Rivers are a source of water for drinking and bathing. Early civilizations were founded along rivers.

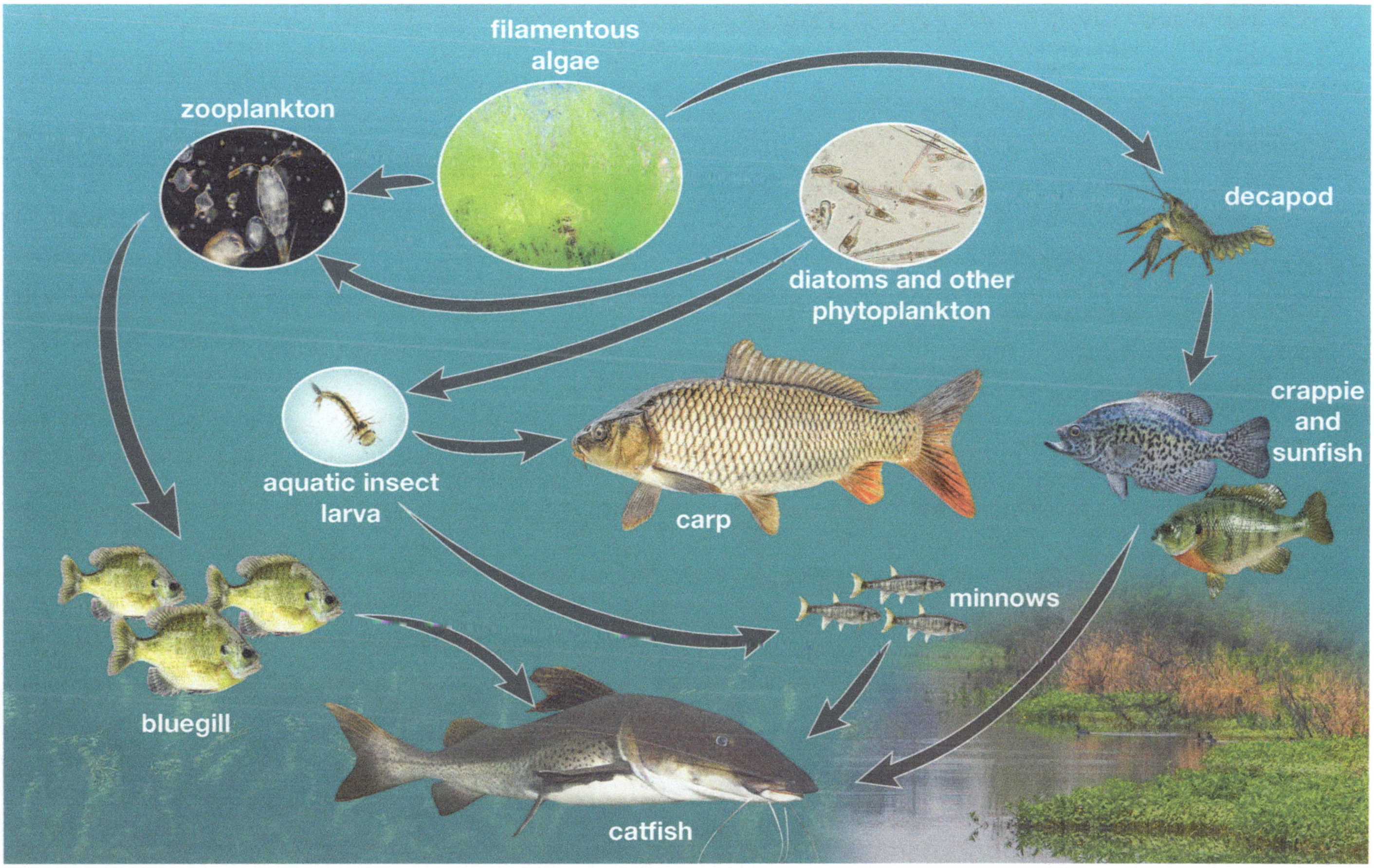

Fig. 4.3 A river food web

- Humans eat fish, crawfish, and other species found in rivers.
- Rivers leave the land next to them filled with nutrients. Farming and agriculture are often done along rivers.
- People use rivers to transport goods on ships.
- Nutrients, dirt, and oxygen make their way through various landscapes and into the world's oceans through rivers.
- Rivers change the shape of the landscape. In fact, as more people decide to settle and work by rivers, flood control measures must be put in place. This is because as the river naturally changes, it might cause harm to the people and businesses nearby.
- Rivers are an important habitat for many different organisms.
- Some rivers power cities. As the water is held back by man-made dams, the energy from the moving water is converted into electrical energy, which can provide electricity for large cities.

Threats to river ecosystems include:

- widespread destruction of riparian zones due to construction, agriculture, or developmental purposes
- pollution from runoff, including sediments from excess fertilizers, pesticides, and toxic chemicals
- dumping of trash directly into rivers by people and businesses
- decreased water flow due to drought or overuse

River ecosystems need to be protected from pollution. Protecting aquatic ecosystems protects biodiversity, keeps habitats that filter water intact, and ensures that food from these ecosystems is safe to eat.

What is an Estuary?

Estuaries are areas where rivers meet the oceans, or where surface fresh water mixes with surface salt water. This water is brackish, which means somewhat or slightly salty. There are many names for estuaries, such as bays, lagoons, inlets, harbors, or sounds. These places where the rivers meet the oceans are an amazing part of nature.

This estuary is where the Miño River in Spain meets the Atlantic Ocean.

Estuaries make up some of the most important ecosystems on the earth. They help maintain water quality by filtering out dirt and other pollutants before they reach the open ocean. Estuaries are also important places for recycling nutrients. Sediment is washed into estuaries from rivers, and these sediments contain nutrients. The bacteria and other organisms within these estuaries then eat the nutrients. These organisms are then eaten by other organisms, and this process continues until the nutrients have cycled through the entire food chain. As organisms die, the nutrients are returned to the sandy bottoms and taken in by plant roots, starting the cycle over again.

Estuary Food Webs and Resources

Estuaries are diverse ecosystems that contain a range of plants, animals, and other

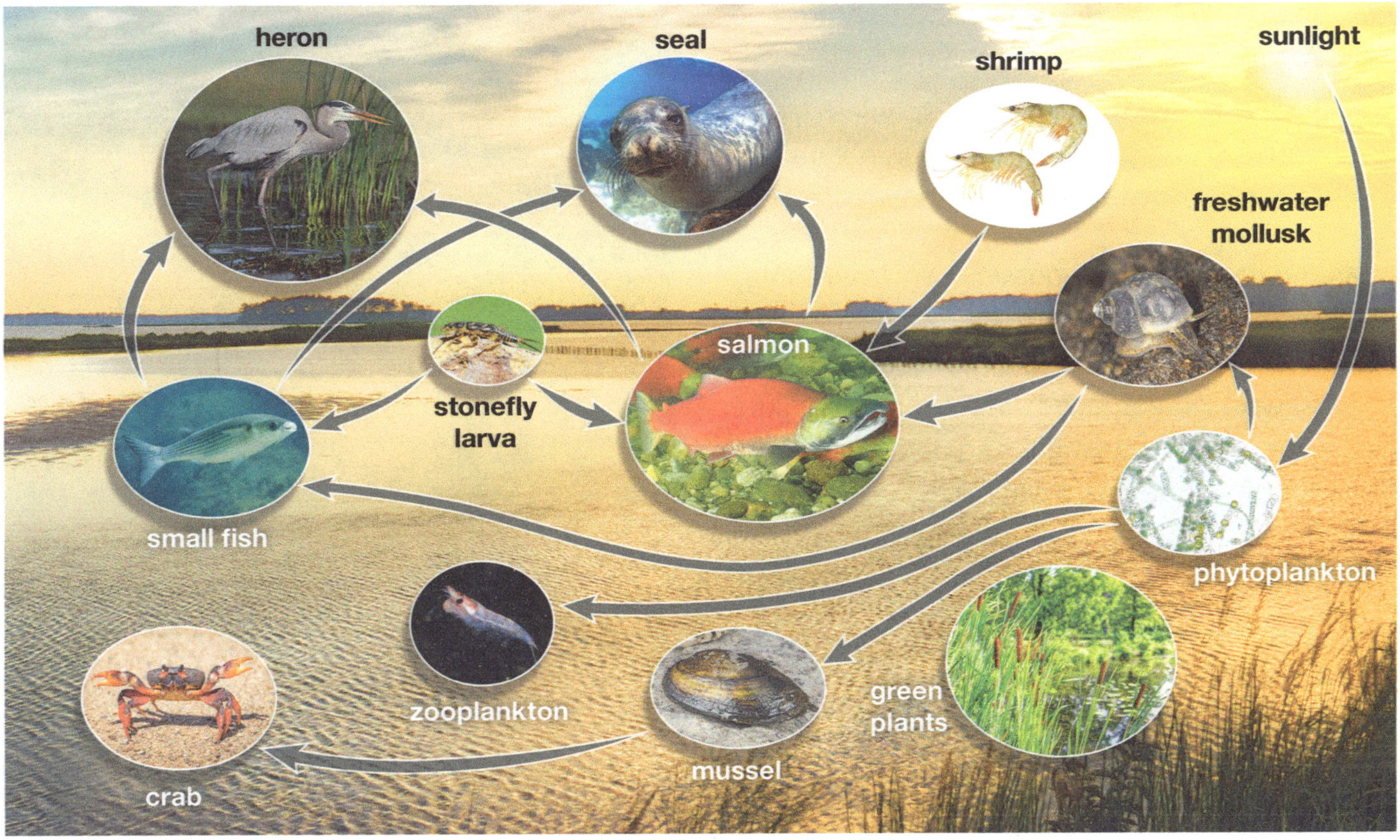

Fig. 4.4 Estuary food web

organisms. They are home to species such as oysters, coral reefs, mangrove trees, manatees, crabs, shrimp, algae, clams, and other types of fish (**Fig. 4.4**).

While some organisms live in estuaries their whole lives (oysters and crabs), others are there for just a part of their life cycle (salmon). Estuaries are full of life, including organisms classified as plankton, nekton, and benthos, as well as aquatic birds and land animals.

Human Interdependence and Impact on Estuaries

A wide variety of human activity is dependent on estuaries, as many people live, play, and work in or near them. People rely on estuaries for activities such as fishing, crabbing, and shipping.

Human activities that can impact estuaries include:

- improper building, which can lead to too many sediments entering estuaries
- direct and indirect exposure to pollution due to trash and increased amounts of fertilizers and pesticides. This pollution can cause unsafe drinking water, algae

blooms, a loss of habitat, and death for many of the aquatic organisms

- the introduction of invasive species resulting in a loss of diversity in native plants and animals

As with most aquatic environments, education and regulation are good ways to start protecting estuaries. A loss or damage to our world's estuaries will eventually hurt humans as well, causing a negative impact on many of the world's ecosystems.

Match the terms with the definitions.

4.4) _____ benthos

4.5) _____ eutrophic lakes

4.6) _____ river

4.7) _____ lake

4.8) _____ nekton

4.9) _____ oligotrophic lakes

4.10) _____ riparian zone

4.11) _____ estuaries

4.12) _____ plankton

A. lakes that tend to be high in nutrients and low in oxygen

B. a still body of surface water deep enough that light cannot reach the bottom

C. lakes that lack nutrients but are full of oxygen

D. microscopic organisms that make up the bottom of the food chain in aquatic ecosystems

E. free-swimming aquatic organisms

F. the area immediately surrounding a lake or river that contains trees, grasses, and other vegetation

G. a stream that contains a large volume of flowing or moving surface water

H. bottom-dwelling aquatic organisms

I. areas where rivers meet the oceans, or where surface fresh water mixes with surface salt water

Choose the best answers using the picture.

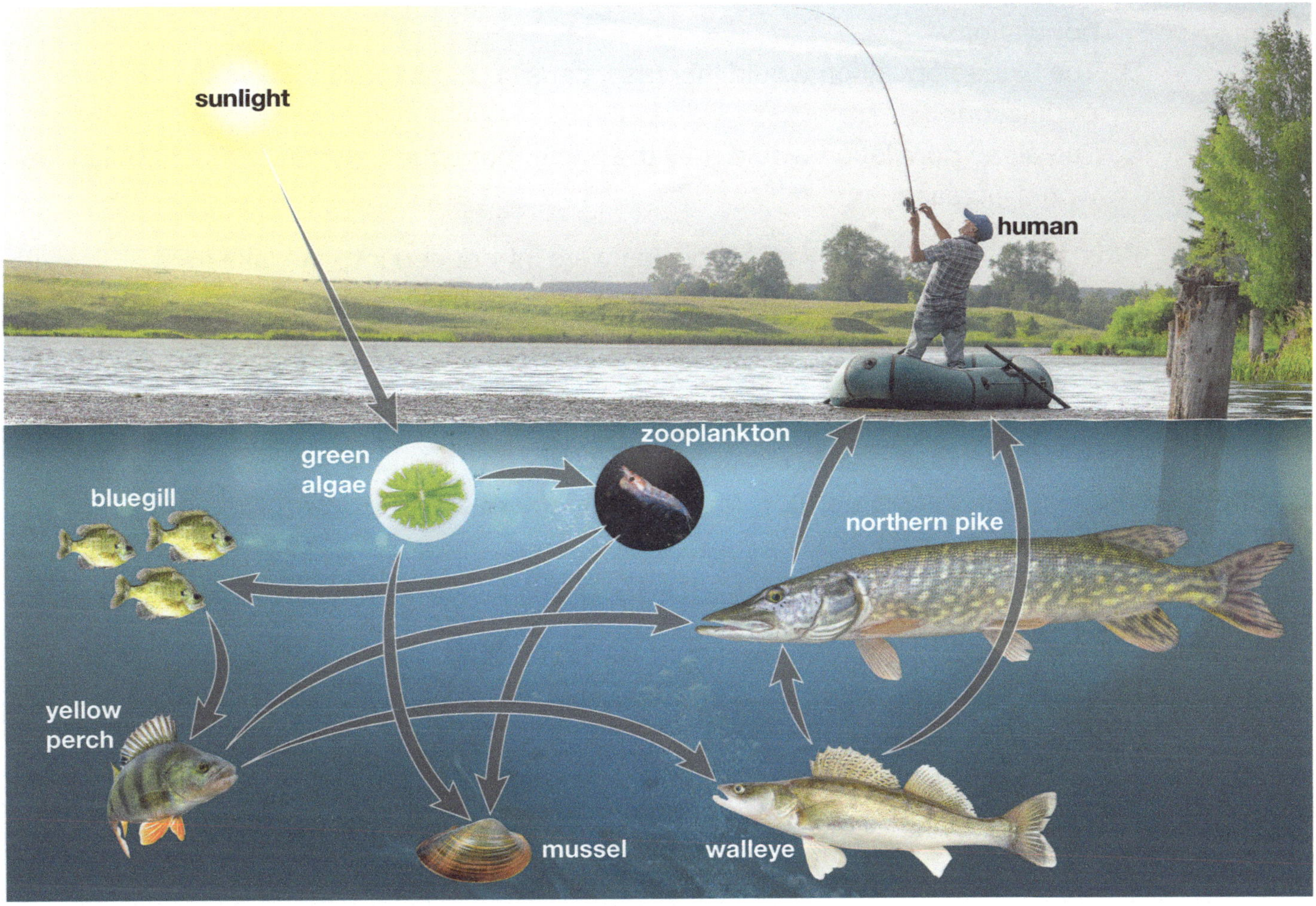

4.13) _____ Which type of organism is the northern pike?
A. plankton
B. benthos
C. nekton

4.14) _____ Which type of organism is the green algae?
A. plankton
B. benthos
C. nekton

4.15) _____ Which type of organism is a mussel?
A. plankton
B. benthos
C. nekton

4.16) _____ What would likely happen if excess fertilizer entered the lake through runoff?

A. The algae population would decrease, causing an increase in oxygen and fish populations.
B. The algae population would increase, causing a decrease in oxygen and fish populations.
C. The algae population would stay the same, but either oxygen or fish populations would increase.

4.17) _____ What would happen if invasive species were introduced into the lake?

A. Oxygen levels would increase and fish populations would increase.
B. Oxygen levels would increase and fish populations would decrease.
C. Oxygen levels would decrease and fish populations would increase.
D. Oxygen levels would decrease and fish populations would decrease.

Match the words with the statements regarding why rivers are important to people.

4.18) _____ transportation
4.19) _____ electricity
4.20) _____ agriculture
4.21) _____ early civilizations

A. Rivers are dammed, converting the energy from moving water.
B. People move goods on ships through rivers.
C. Rivers are a source of water for drinking and bathing.
D. Rivers leave the land next to them filled with nutrients.

Choose the best answers.

4.22) _____ The place where a river meets an ocean is called a(n) ___.

A. coral reef
B. delta
C. estuary
D. island

4.23) _____ The water in an estuary is ___, a mixture of salt water and fresh water.

A. nutrient rich
B. muddy
C. oxygen rich
D. brackish

4.24) _____ *A reason estuaries are important is that they help maintain water quality by ___ dirt and other pollutants before they reach the open ocean. That is why they must be protected from extreme amounts of nutrients in farm runoff and from invasive species.*

A. increasing
B. compacting
C. filtering out
D. cooling

4.25) _____ *Mangrove forests are found in estuaries, which are very salty. Mangrove trees provide a habitat for animals to live in and are a place for many fish species to breed and reproduce.*

What adaptation does a mangrove tree have to survive in an estuary?

A. camouflage

B. contributes to biodiversity

C. filters excess salt

D. mimicry

CHECK ☐ CORRECT ☐ RECHECK ☐

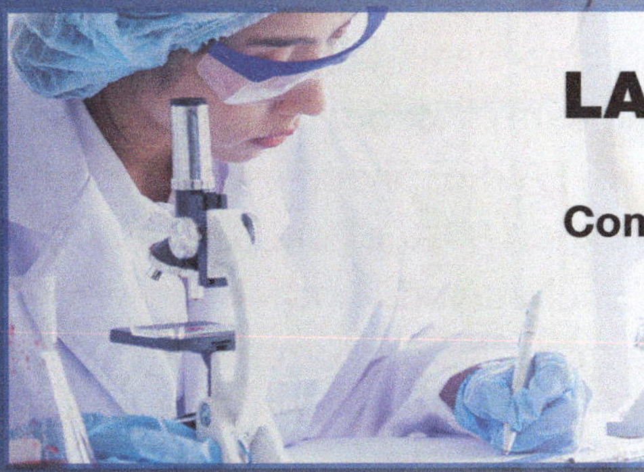

LAB: ECOLOGICAL OBSERVATIONS

Complete this lab in the Lab Manual.

Teacher Initial ☐

CHARACTER SPOTLIGHT

OBSERVANT

The mind can handle only one important task at a time. Some things are very automatic and can be handled with very little conscious effort. However, the truly important tasks require undivided attention and razor-sharp focus. This type of concentration never occurs with a radio playing in headphones or with a phone in one's hand.

If one is to be successful in a career or a relationship, the art of fixed and undivided concentration must be mastered. At the same time, a person must learn to be quiet within so that the true importance of what is being seen, heard, or experienced may be appreciated.

5. MARINE AND SALTWATER ECOSYSTEMS

In the last Lesson, we looked at freshwater ecosystems. In this Lesson, we will take a look at marine or saltwater ecosystems. Marine or saltwater ecosystems are ecosystems that are found in bodies of water that have a high salt content. Most of the earth's water is salt water found in oceans. Even though the water is not drinkable, oceans are very important to people.

THE IMPORTANCE OF OCEANS

There are five oceans on the earth. They are the Atlantic Ocean, the Indian Ocean, the Southern Ocean, the Arctic Ocean, and the Pacific Ocean. Each of these oceans provides a unique landscape and has its own unique wildlife, from the whales and dolphins of the Atlantic to the giant squid of the Pacific. Oceans play a crucial role in regulating the earth's climate and weather patterns. In addition, they have significant economic and cultural importance.

Humans need the oceans for many different reasons. Oceans serve as a source for food, raw materials, energy, transportation, and recreation.

- Every year the ocean provides billions of pounds of fish and other seafood for commercial use. For example, tuna, cod, and shellfish all come from the ocean.
- The ocean provides raw materials for fertilizers through fish and seaweed. Food additives are made from seaweed.
- Many minerals are mined from the ocean.
- Energy from the ocean comes through turbines and offshore drilling that collects oil to be used for gasoline and petroleum products. Tides, waves, and wind over the ocean also provide an energy source to generate electricity.
- The ocean is used for transportation as products are moved across the world and people travel by ferries.
- The ocean helps the climate to be more moderate. Ocean currents bring warm tropical waters to the East Coast of the United States. These tropical waters help warm up the atmosphere in these areas.
- Humans use the ocean to have fun by swimming, surfing, deep-sea fishing, or taking cruises.

The ocean has been called the last great frontier on the earth. According to the National Oceanic and Atmospheric Administration (NOAA), more than 80% of the ocean remains unmapped, unobserved, and unexplored by humans. The materials and organisms at these depths could provide us with the answers to energy problems, new medicine to treat serious illnesses, and a better understanding of how ocean ecosystems function. Learning more about the ocean, which covers about 70% of the earth, offers endless possibilities for new and exciting discoveries.

REVIEW

Choose the best answers.

5.1) _____ Every year the ocean provides ___ of pounds of fish and other seafood for commercial use.

A. hundreds
B. thousands
C. millions
D. billions

5.2) Choose THREE ways that the ocean provides energy for electricity.

☐ energy from the tides and waves
☐ oil from offshore drilling
☐ energy from wind turbines
☐ energy from burning sand

5.3) _____ The ocean provides raw materials for ___ with fish and seaweed.

A. minerals
B. vitamins
C. fertilizers
D. diamonds

5.4) _____ The ocean helps the climate to be more ___ due to currents bringing warm surface water to colder areas.

A. moderate
B. icy
C. extreme
D. dry

5.5) _____ Many ___ are mined from the ocean.

A. minerals
B. vitamins
C. diamonds
D. fertilizers

5.6) _____ According to the National Oceanic Atmospheric Administration (NOAA), how much of the ocean remains unmapped, unobserved, and unexplored by humans?

A. 5%
B. 25%
C. 70%
D. 80%

Open Ocean Resources

The biotic factors within open ocean ecosystems include all the organisms that make up the plankton, nekton, benthos, decomposers, and aquatic birds categories. The abiotic factors include light penetration, nutrients, currents, salinity levels, and temperature.

While there is a wide variety of biotic and abiotic factors in open ocean ecosystems, the open ocean is not a very productive ecosystem. In ecology, **productivity** refers to the rate at which new biomass is produced. **Biomass** is the total mass of an organism, or the total amount of organisms within an ecosystem.

In order for there to be high productivity in an ecosystem, two factors need to be present: sunlight and nutrients. Some parts of the ocean have high levels of nutrients from dead and decaying matter (detritus) but receive no light. These areas do not have productive ecosystems.

Other parts of the open ocean receive light but do not have the nutrients needed to be considered a highly productive ecosystem. In fact, the open ocean has similar productivity levels to that of a desert. In other words, up to 90% of the open ocean is like a desert in relation to the rate in which new biomass is produced. The open ocean waters are not near enough to rivers to receive nutrients washing down from the land, nor are storms powerful enough to stir up the nutrients at the bottom of the ocean floor to provide the surface organisms with the nutrients they need to thrive. Since the open ocean covers most of our planet, we can see just how little productivity occurs there.

While the open ocean can be compared to a desert in regards to productivity levels, life can still be found in both the desert and the open ocean. The ocean can be further broken down into three zone ecosystems.

Photic Zone Ecosystems

The photic zone is the zone that receives light, and where there is light, there is photosynthesis. The most productive part of the photic zone is the first 80 meters of the water column because this region is dominated by phytoplankton. However, due to lack of nutrients, the amount of phytoplankton existing away from the coastline is limited. Nekton can also be found swimming through the photic zone, but they do not add much productivity. The photic zone acts more like a highway from coastline to coastline rather than a habitat.

The open ocean, away from land, is not a productive ecosystem.

Bathyal Zone Ecosystems

The bathyal zone includes areas close to the bottom of the ocean. Within this zone, we can find many benthic organisms, including clams and mollusks, as well as bacteria and other decomposers. Nekton, such as fish, sharks, giant squid, and whales, can also be found grazing along the ocean floor for their food. The bathyal zone is full of nutrients. However, because the ocean at these depths is completely dark, with the exception of the occasional organisms that create their own light, there is not a lot of productivity. In fact, plant life and phytoplankton cannot exist within the bathyal zone because of the lack of light.

Abyssal Zone Ecosystems

The abyssal zone is home to some of the most out-of-this-world organisms. The life within this zone has adapted to great pressure, cold temperatures, and the total

The ocean provides people with electricity, food, work, and fun.

Some deep ocean creatures have bioluminescence, which is the production of light by living organisms.

lack of light. Many fish and other organisms here are blind. The abyssal zone includes the deepest valleys and trenches of the ocean. Here, producers at the bottom of the food chain include archaebacteria, which perform chemosynthesis. **Chemosynthesis** is the process of converting chemicals into an energy form usable by organisms. These chemicals come from hydrothermal vents along the ocean floor. The vents occur along plate boundaries and often release extremely hot temperatures, as well as sulfur and other nutrients that the archaebacteria convert into energy.

Although the open ocean is not the most productive aquatic ecosystem, it is still an amazing, vast expanse of nature with interesting, specially adapted plants, animals, and bacteria.

Archaebacteria are able to do chemosynthesis around these hydrothermal vents in the depths of the ocean.

Choose the best answers.

5.7) _____ The process of converting chemicals into an energy form usable by organisms is called ___.

A. osmosis
B. photosynthesis
C. endocytosis
D. chemosynthesis

5.8) _____ In order for there to be high ___, ecosystems should be high in both nutrients and light.

A. oil reserves
B. productivity
C. temperatures
D. biodiversity

Write B for Bathyal, A for Abyssal, or P for Photic next to the ocean zone descriptions. Some descriptions may have more than one answer.

5.9) _____ lacks light

5.10) _____ receives light

5.11) _____ low in nutrients

5.12) _____ high in nutrients

5.13) _____ archaebacteria with chemosynthesis

5.14) _____ blind organisms

CHECK ☐ CORRECT ☐ RECHECK ☐

6. OCEAN POLLUTION

As we saw in the last Lesson, oceans are important places that people depend on. People also play a role in polluting the ocean. The main ways that oceans can be polluted are through garbage and oil spills.

GARBAGE

There are two gigantic garbage patches swirling around in the Pacific Ocean. Some estimate their size to be twice the size of Texas, but since most of the garbage lies slightly underneath the water's surface, it is difficult to get accurate estimates. The Pacific garbage patches are caught in rotating currents in the Pacific Ocean (**Fig. 6.1**).

Much of the plastic from garbage in the ocean breaks down into microplastics. Also referred to as mermaid tears or nurdles, microplastics are created when the sun breaks down the plastics into smaller and smaller pieces. They are less than five millimeters long and can be eaten by all aquatic life, even plankton. If filter feeders eat plastic, it can cause them body damage. Other animals that eat the plastic can be poisoned or develop deadly blockages in their digestive tracts.

Microplastics in the ocean are a growing concern because they can also soak up toxic chemicals in the surrounding water. Once one animal eats plastic, biomagnification of the toxins then threatens the whole food web. **Biomagnification** is how the toxin both moves and increases in concentration through each energy level in a food web. The higher an organism is on a food chain, the greater the concentrations of toxins the organism usually has (**Fig. 6.2**). Let's look at methylmercury as an example of both bioaccumulation and biomagnification.

When mercury enters aquatic ecosystems, phytoplankton (algae) ingest the mercury. Everything that eats phytoplankton also gets the mercury, and so on up the food chain. The higher organisms are on the food chain, the

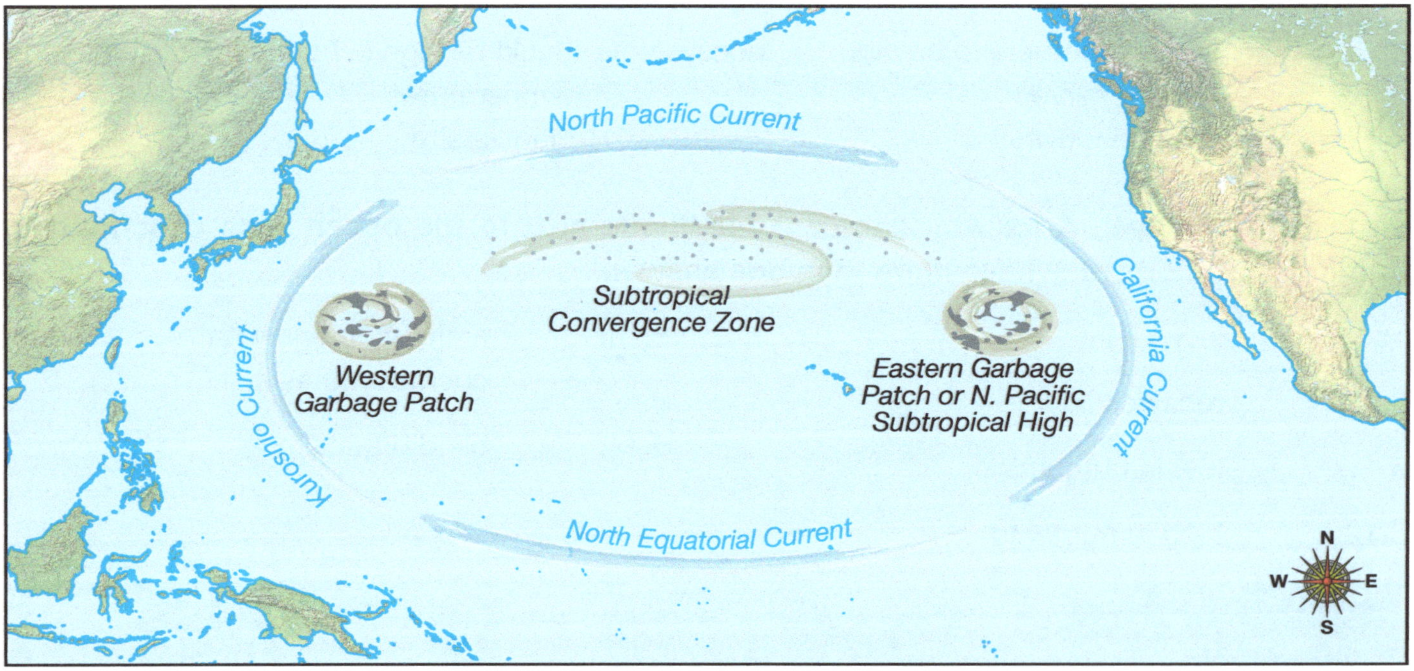

Fig. 6.1 Garbage patches in the Pacific Ocean

greater the mercury concentrations are due to the amounts of food they need to eat to survive.

Toxins like methylmercury can cause birth defects, cancer, and reproductive and growth problems. The main way people are exposed to these types of toxins is through eating the plants and animals that have accumulated the toxin.

Plastics can also biodegrade, releasing chemicals such as bisphenol A (BPA) into the water. **Biodegradable** plastics are those that can be broken down by living organisms into different chemical compounds. It is estimated that 44% of seabirds eat plastic by mistake. Some marine animals eat plastic bags, perhaps mistaking them for jellyfish. Cleaning up trash in oceans is difficult because removing the trash also removes phytoplankton. When phytoplankton are removed, the bottom of the food chain disappears.

Microplastics are created as the sun breaks down plastic into smaller sizes.

Approximately 80% of trash in the ocean originally comes from land. Some coastal communities dump their garbage directly into the ocean. Garbage also enters the ocean from polluted runoff in streams. The key to controlling garbage in the ocean is to have good waste management plans on land and reduce the amount of garbage that enters runoff.

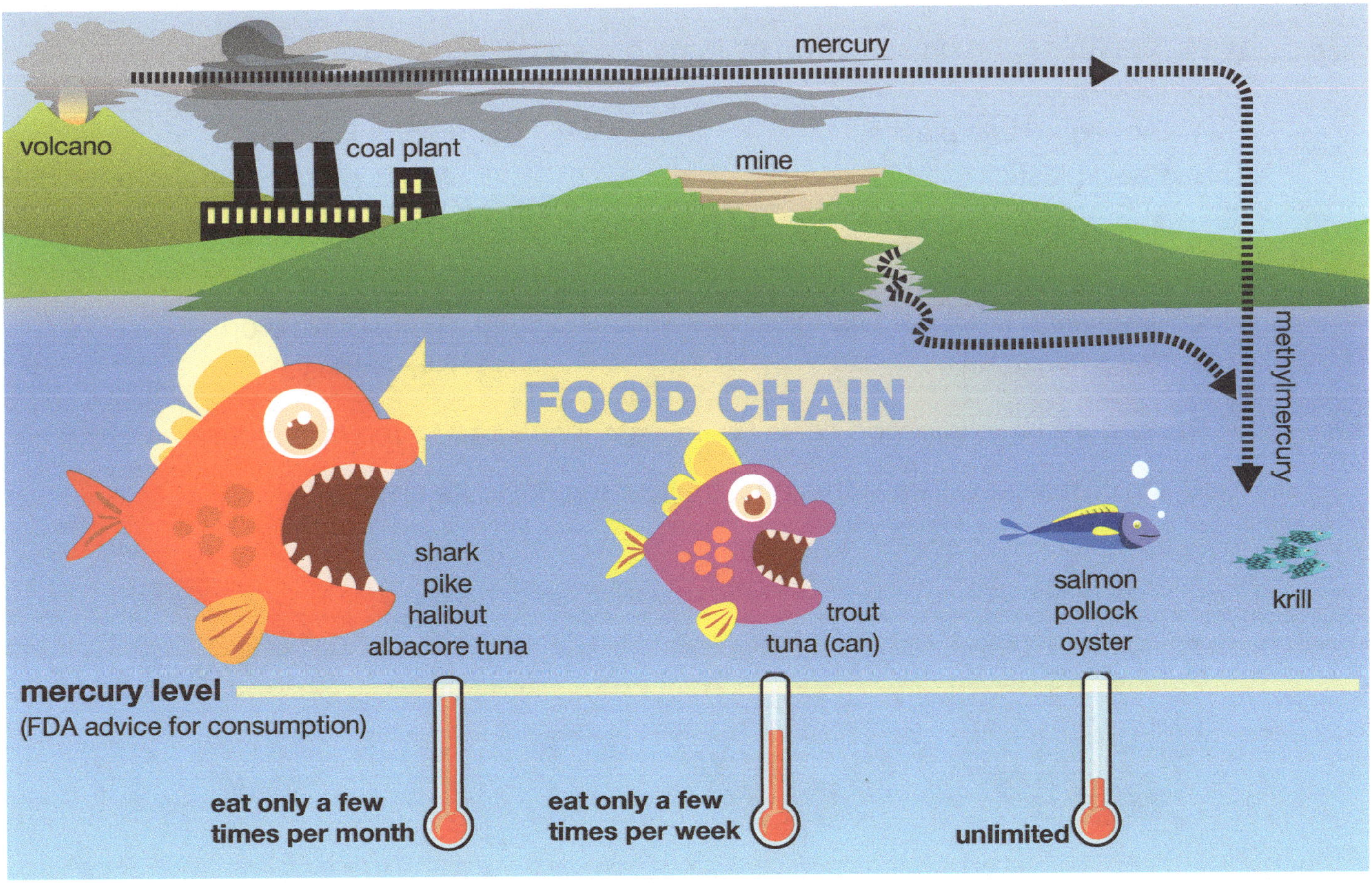

Fig. 6.2 Biomagnification increases toxin levels. Toxin levels are higher in organisms that are higher on the food chain.

REVIEW

Write T for True or F for False. If False, rewrite the statement to make it True.

6.1) _____ It is difficult to get accurate estimates of the size of the garbage patches in the ocean because most of the garbage lies beneath the water's surface.

____

____

6.2) _____ The garbage patches are caught in rotating currents in the Atlantic Ocean.

____

____

6.3) _____ Microplastics are only a danger to larger organisms, like fish.

____

____

Choose the best answers.

6.4) _____ ___ are able to be broken down by living organisms into different chemical compounds.

A. Biodegradable plastics
B. Basic plastics
C. Microplastics
D. Acidic plastics

6.5) _____ ___ is how a toxin both moves and increases in concentration through each energy level in a food web.

A. Biotoxin
B. Bioremediation
C. Biomagnification
D. Biosynthesis

6.6) _____ *The organisms in this food chain live in waters polluted by methylmercury.*

Which organism will have the most methylmercury in its system?

6.7) _____ Oysters have been affected by an oil spill. What might happen to oyster population levels?

A. Population levels will increase because oil has more nutrients.

B. Population levels will decrease because oil can keep oysters from reproducing.

C. Population levels will stay the same because oil has no effect on oysters.

OIL SPILLS

Oil spills are one of the largest human-caused disasters in the ocean. Major oil spills occur in the ocean, but oil can also spill on land and then make its way to the ocean through runoff. Oil spills pollute ecosystems, deplete fisheries, and harm the economy for those who make a living from the ocean.

Oil spills harm wildlife.

Let's take a look at the different causes of oil spills, the consequences, and the solutions for cleaning them up.

Causes

Oil gets in the ocean through several ways.

- Natural seepage
  Oil is formed in the ocean when plankton and other organisms die and are squished together with pressure for a long period of time. These oil reserves have natural leaks and are the number one way that oil spills into the ocean.
- Runoff and human activity
  Polluted runoff and leaks from boats and other equipment account for 85% of oil pollution in the United States.
- Transportation
  Oil leaks from tankers and pipes, and it also enters the ocean through intentional dumping.
- Oil extraction
  As companies drill for oil, things can go wrong. Spills come from offshore platforms and blowouts during the exploration of oil deposits, as well as from the extraction of the oil. The Gulf of Mexico oil spill in 2010 was caused by extraction. An estimated 172 million gallons of oil entered the water over a period of several months.

Damage

Oil spills have much short-term impact on populations of many types of species. Organisms that do not die from an oil spill might be unable to reproduce.

The damage from oil spills is both long- and short-term. Studies show that the short-term effects of oil spills are the most damaging. Wildlife that become soaked in oil are more prone to die immediately after a spill. Even if animals are successfully cleaned up, they often are unable to reproduce. Fish and shellfish also become contaminated with oil and are unable to be harvested, sold, and eaten. Large die-offs of plants and animals happen immediately after an oil spill, as the most harmful chemical compounds break down rather rapidly.

Clean Up

There are several different approaches to cleaning oil out of the water.

- Leave the oil alone and let it break down by natural processes.
- Contain the oil spill and remove oil from the water using skimmer equipment or controlled burning.
- Use agents called dispersants to help spread out the oil so it can be broken down faster.
- Introduce bacteria and other living organisms that essentially eat the oil, helping it break down faster.

Deciding how to clean oil spills depends on the size and location of the oil spill. Sometimes skimming and burning can cause more harm than the actual oil. Bioremediation is rather successful but is not rapid enough for extremely large spills.

Case Study: Ocean Acidification

Many scientists are studying the acidification of the ocean, or how the ocean is becoming more acidic. The ocean is becoming more acidic because of carbon dioxide. Carbon dioxide is a gas that enters the atmosphere when oil, coal, or natural gas is burned. The water in the ocean can absorb the carbon dioxide that is in the air. The more carbon dioxide that is in the air, the more carbon dioxide the ocean absorbs. According to the EPA, the ocean is now 30% more acidic than it was during the Industrial Revolution.

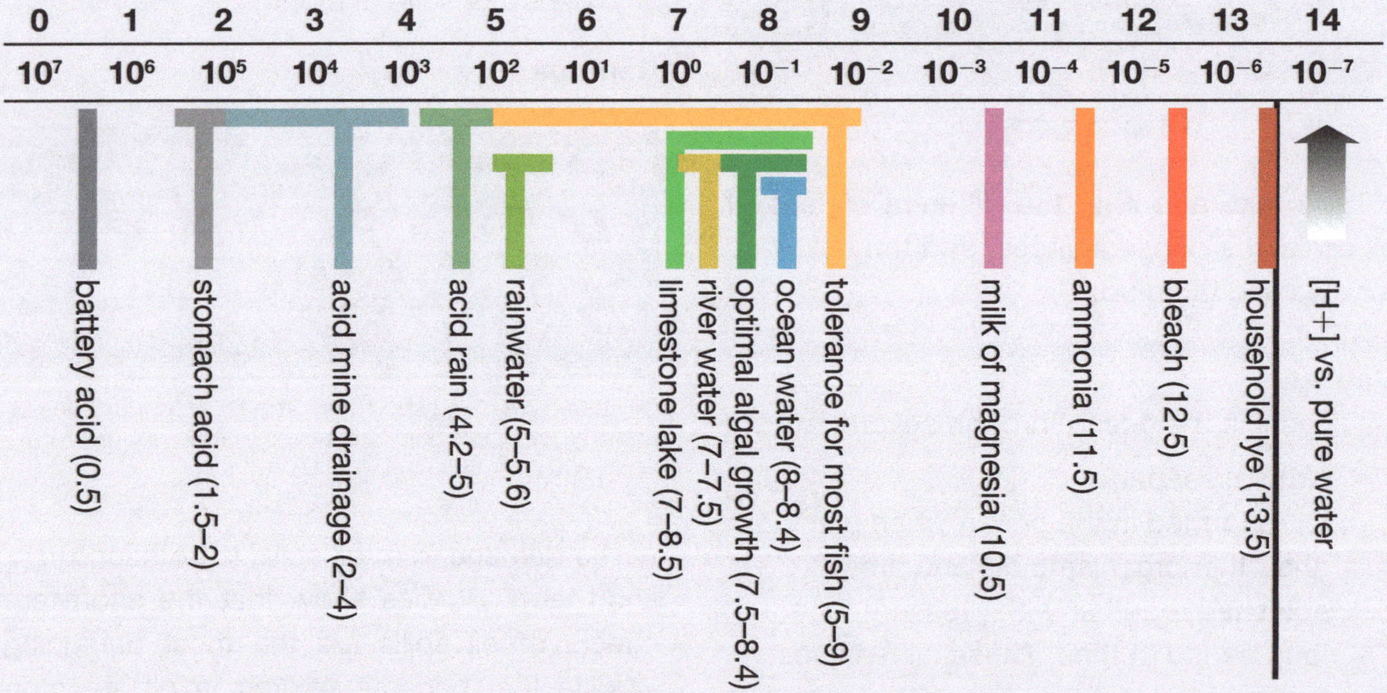

The more acidic water raises water temperature and can cause shells to grow slowly. Extremely acidic water can cause coral to die and shells to be deformed.

Remember that acidity is measured on the pH scale from 0–14. Seven is neutral, greater than seven is called basic or alkaline, and less than seven is acidic. The ocean is about 8.1 on the pH scale, making the water basic. However, it used to be 8.2. Since the pH number has lowered, the ocean is becoming more acidic. The way organisms interact with water pH is a biotic to abiotic interaction.

Write T for True or F for False. If False, rewrite the statement to make it True.

6.8) _____ Ocean acidification can cause water temperature to rise, coral to die, and shells to be deformed.

____

____

6.9) _____ The way organisms interact with water pH is a biotic to biotic interaction.

____

____

6.10) _____ Oil spills have no effect on people.

____

____

Choose the best answers.

6.11) What are FOUR ways oil gets into the ocean?

- ☐ natural seepage
- ☐ precipitation
- ☐ runoff and human activity
- ☐ oil extraction
- ☐ dredging of sediments
- ☐ transportation

6.12) _____ *Carbon dioxide, a greenhouse gas, dissolves in ocean water. As more carbon dioxide is absorbed into the ocean, the water temperature rises and the acidity of the water increases. As the acidity increases, it can dissolve shells on animals, making the shells weaker or more difficult to grow. Crustaceans are a type of animal that has a shell.*

Looking at the following food chain, which statement best describes organisms and ocean acidification?

A. Ocean acidification can potentially cause only crustacean populations to decrease.
B. Ocean acidification can potentially cause crustacean and fish populations to decrease.
C. Ocean acidifications can potentially cause crustacean, fish, dolphin, and killer whale populations to decrease.
D. Ocean acidification can potentially cause only crustacean populations to increase.

CHECK ☐ CORRECT ☐ RECHECK ☐

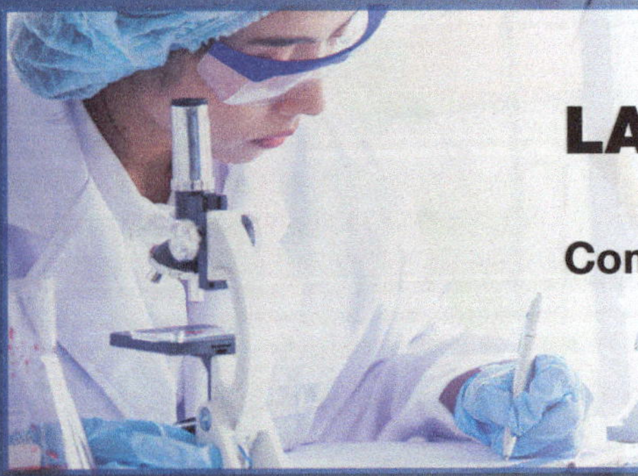

LAB: HUMANS AND POLLUTION

Complete this lab in the Lab Manual.

Teacher Initial ☐

MY SCORE

QUIZ 2

TEACHER INITIAL

(Each question, 4 points)

Fill in the blanks using words from the box below.

benthos	oligotrophic	nekton	estuaries	eutrophic
plankton	riparian zone	lake	river	

2.01) Lakes that tend to be high in nutrients and low in oxygen are considered ________________________.

2.02) Lakes that lack nutrients but are full of oxygen are considered ________________________.

2.03) ________________________ are bottom-dwelling aquatic organisms.

2.04) ________________________ are free-swimming aquatic organisms.

2.05) ________________________ are microscopic organisms that make up the bottom of the food chain in aquatic ecosystems.

2.06) A(n) ________________________ is a still body of surface water deep enough that light cannot reach the bottom.

2.07) A(n) ________________________ is a stream that contains a large volume of moving surface water.

2.08) The area immediately surrounding a lake or river that contains trees, grasses, and other vegetation is known as the ____________________________________.

2.09) ________________________ are areas where rivers meet the oceans, or where surface fresh water mixes with surface salt water.

Choose the best answers.

2.010) _____ As water temperatures increase, oxygen levels ___.

A. increase
B. decrease
C. stay the same

2.011) _____ *Mangrove forests are found in estuaries.*

What adaptation does a mangrove tree have to survive in brackish water?

A. camouflage
B. contributing to biodiversity
C. mimicry
D. filtering excess salt

2.012) Rivers are important because of which THREE things?

- ☐ Goods can be transported on ships.
- ☐ The land next to rivers is filled with nutrients.
- ☐ There is water available for drinking and bathing.
- ☐ They increase sediments that reduce oxygen production.

2.013) _____ The vegetated area next to the river is called the ___.

A. vegetation zone
B. riparian zone
C. aquatic zone
D. bathyal zone

2.014) _____ Which type of ecosystem will you find if you follow the river to the place where it meets the ocean?

A. estuary
B. lake
C. pond
D. creek

2.015) Which THREE things could cause oxygen levels to decrease in estuaries?

- ☐ reducing the amount of nutrients entering the runoff from farms
- ☐ increasing the amount of nutrients entering the runoff from farms
- ☐ finding native species to filter pollutants
- ☐ introducing an invasive species
- ☐ removing trees from riparian zones

2.016) _____ *The silver carp is an invasive species in many rivers in the United States. Silver carp and steelhead compete for the same food resources.*

What would happen to the steelhead populations if silver carp were introduced into its ecosystem?

A. Population levels would increase.
B. Population levels would decrease.
C. Population levels would stay the same.

2.017) _____ ___ is the process of converting chemicals into any energy form usable by organisms.

A. Chemosynthesis
B. Endocytosis
C. Photosynthesis
D. Osmosis

2.018) Choose THREE things that help the ocean provide energy for electricity.

- ☐ energy from the tides and waves
- ☐ oil from offshore drilling
- ☐ energy from wind turbines
- ☐ power generated by electric eels

2.019) _____ The ocean helps the climate to be more moderate due to the currents bringing warm surface water to ___ areas.

A. colder
B. dry
C. warmer

2.020) _____ According to the ___, more than 80% of the ocean remains unexplored.

A. National Oceanic and Atmospheric Administration
B. United States Coast Guard
C. Australian Marine Conservation Society
D. Ocean Conservancy

2.021) _____ In order for there to be high productivity, ecosystems should be high in both nutrients and ___.

A. temperatures
B. sunlight
C. salinity
D. biodiversity

Choose the best answers.

2.022) _____ *The organisms in this food chain live in waters polluted by methylmercury.*

Select the organism that will have the most methylmercury in its system.

2.023) _____ *Oysters have been affected by an oil spill.*

What might happen to oyster population levels?

A. Population levels will increase because oil has more nutrients.
B. Population levels will decrease because oil can keep oysters from reproducing.
C. Population levels will stay the same because oil has no effect on oysters.

2.024) _____ *Carbon dioxide, a greenhouse gas, dissolves in ocean water. As more carbon dioxide is absorbed into the ocean, the acidity and temperature of the water increases. As the acidity increases, it can dissolve shells on animals, making the shells weaker or more difficult to grow. Crustaceans are a type of animal that has a shell.*

Looking at the following food chain, which statement best describes organisms and ocean acidification?

A. Ocean acidification can potentially cause only crustacean populations to decrease.
B. Ocean acidification can potentially cause crustacean and fish populations to decrease.
C. Ocean acidifications can potentially cause crustacean, fish, dolphin, and killer whale populations to decrease.
D. Ocean acidification can potentially cause only crustacean populations to increase.

2.025) Which FOUR ways does oil get into the ocean?

- ☐ natural seepage
- ☐ precipitation
- ☐ runoff and human activity
- ☐ oil extraction
- ☐ dredging of sediments
- ☐ transportation

CHECK ☐ CORRECT ☐ RECHECK ☐

SECTION THREE

Objectives:

- Understand the role that scale, proportion, and quantity play in ecosystem structure and performance.
- Identify key challenges that exist when trying to predict and manage change within an ecosystem.
- Explain the role of engineering in addressing and reducing the impacts of ecosystem changes.
- Understand how engineers help protect and restore ecosystems.
- Explain the role of an environmental engineer.
- Describe the contributions of Dr. Ellen Swallow Richards.
- Identify the impact of Dr. Ellen Swallow Richards in the field of environmental engineering.

Vocabulary:

proportion *[pruh-PAWR-shuhn]* – the relative numbers of different organisms within an ecosystem

quantity – the amount of resources available in an ecosystem

scale – the size of the different parts of an ecosystem

7. PREDICTING ECOSYSTEM CHANGES

Both land and marine ecosystems are incredibly complex. In land ecosystems, factors such as climate, topography, soil type, and the presence of different species all interact to determine the types of plants and animals that can survive in a particular area. The relationships between these living organisms, as well as the interactions between living organisms and their physical environment, can be intricate and delicate. For example, one species might depend on another for food or pollination, while changes in temperature or rainfall can cause ripple effects throughout the ecosystem. Human activities such as deforestation, pollution, and the introduction of invasive species can also have major impacts on land ecosystems. These changes can alter the balance of the ecosystem, leading to declines in biodiversity.

Coral reefs have the highest biodiversity of any ecosystem on the planet.

Marine ecosystems are just as complicated as land ecosystems. Different factors like temperature, salinity, light, and water flow play an important role in determining which creatures can live in a certain area. In the ocean, different organisms live at different depths and areas of the ocean. They also each depend on each other for food and shelter.

Coral reefs are an important example. Coral reefs are considered the most diverse ecosystems on the planet. They are home to many different species, including fish, turtles, and sharks. Coral reefs help protect coastlines from erosion and big waves. They also provide economic benefits by being a place for fishing and tourism. Corals are good gauges of the health of the ocean because they are sensitive to changes in their environment. Healthy corals depend on stable ocean temperatures, which can be easily disrupted by human activity.

Mass deforestation is wiping out species of mammals all around the world.

The condition and function of both land and marine ecosystems are influenced by a variety of factors. However, ecosystem health and stability rely heavily on diversity, which is closely tied to the concepts of scale, proportion, and quantity.

Scale, Proportion, and Quantity

Diversity in an ecosystem refers to the variety of different species that live and interact within a specific environment. It is important because it helps make the ecosystem strong and better able to survive changes. Think about a big puzzle with many different pieces. Each piece is important and helps the puzzle stay together. In the same way, each organism in an ecosystem is important and helps the ecosystem work properly.

Scale

When we talk about scale, we usually mean the size of something. In ecology, **scale** refers to the size of the different parts of an ecosystem and how they relate to each other. For example, if you have a small garden, the scale of the plants and animals in it would be different than if you had a big, sprawling forest.

In a small ecosystem, like a garden, there may be fewer predators and fewer prey animals. The predator population may be limited by the availability of food (prey), and the prey population may be limited by the presence of predators. If there are more predators, then the prey population may decrease. This can also cause the predator population to decrease because they have less food to eat.

In a larger ecosystem, like a forest, there may be more predators and more prey animals. The predator population may be able to sustain itself even if there are changes in the availability of prey. In a smaller ecosystem, the relationship between predator and prey populations may be more fragile and easily disrupted. In larger ecosystems, the relationship may be more stable and resilient.

The increase or decrease of an ecosystem area can affect predators and prey by changing the amount of space and resources available to them. This can cause their poulations to increase or decrease. For instance, let's say that there is a forest with lots of deer and some wolves. If people were to come and cut down parts of the forest, the deer would have less space to move around and less food to eat. This would cause the deer population to decrease, which means the wolves would have less food to eat and their population would also decrease.

If the number of predators decreases because there is not enough food for the predators to eat, they will have to move to another area to find food. A decrease in predators can then lead to an increase in the population of their prey, which can upset the balance of the ecosystem.

PROPORTION

Proportion is the relationship between the different parts. More specifically, it refers to the relative number of different organisms within an ecosystem. On a small scale, a small change in the proportion or number of a particular species can have a large impact on the rest of the ecosystem. For example, the loss of a keystone species can result in a ripple effect that moves through the ecosystem and causes major changes to the population sizes and interactions of other species. For example, if a keystone species like the sea star, which helps control the population of mussels in a marine ecosystem, were removed, the mussels would begin to overgrow and outcompete other species for resources. This could cause a decline in the diversity of species in the ecosystem.

Overgrown mussels take over coastal rocks.

The proportion of predator and prey populations is also important in maintaining the balance of an ecosystem. For example, if the number of lions in a savanna decreases, the proportion of lions to gazelles will also decrease, which can result in an overpopulation of gazelles. This can have a negative impact on the savanna ecosystem.

QUANTITY

Quantity refers to the amount or number of something. In an ecosystem, it is important to have the right quantity of different types of plants, animals, and other resources to keep everything in balance. A change in the quantity of a resource, such as water or food, can have a big impact on the organisms that depend on it. For example, increased rainfall can lead to increased plant growth, which will increase the food supply for herbivores and benefit the entire food web. An increase in the quantity of a resource can lead to greater diversity, while a decrease can result in less diversity within an ecosystem.

The number of predator and prey populations is directly related to the ecosystem's balance. If the quantity of predators decreases, it can lead to an increase in the quantity of prey, and vice versa. For example, if the number of shark populations decreases in an ocean ecosystem, the quantity of smaller fish may increase. This can have a negative impact on the ecosystem, as the increased number of smaller fish may lead to a shortage of food for other species, such as larger fish or sea birds.

Large school of fish swimming along a coral reef

Table 7.1 shows examples of how changes in scale, proportion, and quantity can influence the relationship between predator and prey populations and the balance of an ecosystem. An imbalance in any of these factors can cause disruptions that may lead to ecological collapse. Understanding these relationships helps us better understand how ecosystems work and how we can help maintain a healthy and stable environment for all species.

TABLE 7.1 ECOSYSTEM INTERACTIONS

Factor	Increase in Population	Decrease in Population
Scale	If the ecosystem area increases (e.g., brush growth after rainfall), the populations of prey and predators may also increase.	If the ecosystem area decreases (e.g., brush loss after drought), the populations of prey and predators will decrease.
Proportion	If the number of predators (e.g., lions) increases, the proportion of predators to prey (e.g., gazelle) will increase.	If the number of prey (e.g., deer) increases, the proportion of predators to prey will decrease.
Quantity	If the quantity of predators (e.g., sharks and dolphins) increases, the quantity of prey (e.g., smaller fish) may decrease.	If the quantity of prey (e.g., zooplankton) decreases, the quantity of predators (e.g., whales) may also decrease.

REVIEW

Choose the best answers.

7.1) _____ Which of the following best describes the difference between scale, proportion, and quantity in an ecosystem?

A. Scale refers to the size of the ecosystem; proportion refers to the relative number of different organisms within the ecosystem; and quantity refers to the amount of a particular resource or species within the ecosystem.

B. Scale refers to the amount of time it takes for changes to occur within the ecosystem; proportion refers to the distribution of different species within the ecosystem; and quantity refers to the size of the ecosystem.

C. Scale refers to the amount of resources available within the ecosystem; proportion refers to the relationship between different components within the ecosystem; and quantity refers to the number of organisms within the ecosystem.

D. Scale refers to the geographical area an ecosystem covers; proportion refers to the amount of human impact on an ecosystem; and quantity refers to the number of different species within an ecosystem.

7.2) Which TWO of the following describes the potential impact of changes in the population size on an ecosystem?

☐ A decrease in the population size of a predator will have no impact on the ecosystem.

☐ A decrease in the population size of a predator will lead to an increase in the population size of its prey.

☐ An increase in the population size of a prey will lead to an increase in the population size of its predator.

☐ An increase in the population size of a prey will lead to a decrease in the population size of its predator.

7.3) _____ What would happen to the food web of an ecosystem if the proportion of a certain species of fish were to significantly decrease?

A. The population of other species that rely on that fish for food would also decrease.

B. The population of other species that rely on that fish for food would increase.

C. There would be no change to the food web.

7.4) _____ What happens to the ecosystem if the quantity of a key resource, such as water or food, decreases?

A. The ecosystem becomes more diverse.

B. The ecosystem becomes less diverse.

C. There is no change in the ecosystem.

D. The ecosystem becomes more stable.

7.5) _____ What would happen to the food web of an ecosystem if the proportion of two species changes?

A. It can cause a ripple effect on the entire food web.
B. It will not affect the food web.
C. It will only affect the population of one species.
D. It will only affect the organisms of two species.

Predicting and Managing Change

Predicting and managing changes in an ecosystem are important for protecting biodiversity, maintaining ecosystem balance, and minimizing human impact on the environment. Ecosystems provide many important resources, such as food, water, and medicine that are essential for our survival.

However, understanding and managing changes within an ecosystem can be hard because ecosystems are complex systems made up of many different parts all working together. These complexities can make it difficult to identify which changes to the ecosystem should be managed and in what ways. Some of the challenges scientists face when trying to predict and manage change within an ecosystem include the following:

- Complexity
  Ecosystems are made up of a wide variety of interacting components, including living organisms and the nonliving environment. The interactions between these components can be complex. It is hard to know how changes in one part will affect other parts.
- Uncertainty
  It is difficult to know what will happen in the future. Factors such as climate change, population growth, and land use can have a significant impact on ecosystems; however, it can be hard to predict the exact nature and extent of these impacts.
- Nonlinearity
  Many of the changes that occur within an ecosystem are known to have a nonlinear effect on the environment. This means that the impact of a change does not always match its size. Small changes can often have big and surprising effects. Because of this, it is difficult to know how the ecosystem will respond to different changes.

Climate change contributes to droughts that are more frequent, longer, and more severe.

- Scale
  Ecosystems can range in size from small ponds to large regions of land or sea. Managing changes within an ecosystem can be challenging because scientists need to adjust their approach depending on its size.
- Human impact
  Many of the changes that occur in ecosystems are caused by human activities. It is sometimes hard to separate natural changes from human-made changes. Understanding how human activities affect ecosystems is important to managing them.

Despite the complex nature of ecosystems, understanding and managing changes within them is vital for ensuring the survival of the essential resources they provide.

Structure-Function Relationship

Let's imagine that you're in a big park with a lot of different plants and animals. Each of these plants and animals has a specific shape, or structure, that helps them do the things they need to do to survive. Structure and function are interdependent. This is called the structure-function relationship.

For example, a bird has a beak that is just the right shape for cracking open nuts and seeds to eat. The shape of the beak determines what the bird can eat and how it eats.

Now, there are also a lot of different things that can affect the structure or function of plants and animals in the park, like the weather, the amount of food and water available, and the other plants and animals that live there. These things can change, and when they do, it can affect the plants and animals in the park.

For example, if it starts raining a lot, the plants might start growing taller, and there might be more food for the animals. But if it rains too much, there could be a flood, and some of the plants and animals might have trouble surviving.

So, you can see how the structure and function of objects in an ecosystem, like the plants and animals in the park, are related and how changes in the environment can impact the stability and change of these objects.

Choose the best answers.

7.6) Which FOUR are some major challenges of predicting and managing change within an ecosystem?

- ☐ complexity
- ☐ certainty
- ☐ nonlinearity
- ☐ scale
- ☐ human impact

7.7) _____ What is one consequence of nonlinearity in an ecosystem?
A. Small changes lead to big and unexpected effects.
B. Large changes lead to small and predicted effects.

7.8) _____ Which of the following is an example of the structure-function relationship in a park ecosystem?
A. the growth of plants
B. the beak of a bird
C. the amount of food available
D. the weather patterns

7.9) Which TWO of the following best describes the relationship between structure and function of objects in an ecosystem?
☐ The structure of an object does not affect its function.
☐ The function of an object does not affect its structure.
☐ The structure and function of an object are unrelated.
☐ The structure and function of an object are interdependent.
☐ Changes in the environment can impact the stability and change in objects.

7.10) _____ Which of the following environmental factors would most likely impact the stability and change of objects in a park ecosystem?
A. the shape of a bird's beak
B. the number of food web levels
C. the weather patterns
D. the study of the park's history

CHECK ☐ CORRECT ☐ RECHECK ☐

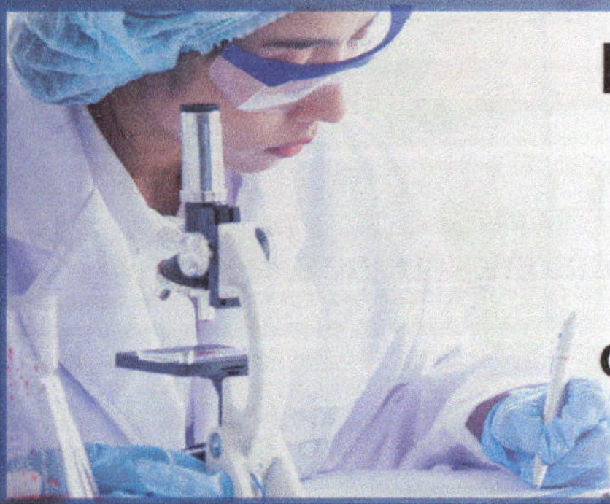

LAB: IMPACT OF SCALE, PROPORTION, AND QUANTITY ON AN AQUATIC ECOSYSTEM

Complete this lab in the Lab Manual.

Teacher Initial ☐

8. ENGINEERING SOLUTIONS FOR ECOSYSTEM CHANGES

Ecosystems change all the time because of things like weather, natural disasters, and the evolution of different animal and plant species. These changes can have both good and bad effects on the plants, animals, and other living things within an ecosystem. Human activities, such as deforestation, pollution, and climate change, can also have a significant influence on ecosystems.

Engineering Solutions for Ecosystem Changes

Engineers are people who design and build things to help protect and preserve the environment. They use their knowledge of science and technology to come up with solutions to problems caused by changes in ecosystems. Here are a few examples of the types of things they do:

- Climate change adaptation
  Engineers create systems and technologies that help communities adapt to changes in the weather, such as rising sea levels or more extreme storms. For example, they might make irrigation systems that can survive droughts or design buildings that can withstand strong winds.
- Ecosystem restoration
  Engineers work with groups that want to restore damaged ecosystems or protect ecosystems that are in danger. They might help design and build dams or levees to control the flow of water in an ecosystem or plan for reforestation (planting trees) or habitat restoration (making sure animals have a place to live).
- Water management
  Engineers design and build systems that help control the flow of water in an ecosystem. For example, they might make dams or reservoirs that store water or treatment plants that clean the water. These systems can also help conserve and protect water sources, increase access to clean drinking water, improve agricultural productivity, and create a dependable source of renewable energy.

Flood defense strengthened in New Orleans after Hurricane Katrina

- Sustainable agriculture
  Engineers help farmers use less water and other resources by developing precision irrigation systems. These systems use technology to apply the exact amount of water, fertilizer, and other nutrients to crops in the right place at the right time.
- Environmental monitoring
  Engineers create tools and techniques to monitor the health of ecosystems. For example, they might use sensors and monitoring networks that track water quality, air quality, or wildlife populations.

These are just a few examples of the many things engineers do to help protect and preserve the environment. By using science and technology, they work to minimize the negative impacts of ecosystem changes and help keep ecosystems healthy and resilient.

Irrigation system watering field of greens

Methods and Tools

Scientists use different tools to study and learn about ecosystems. Here are some examples:

- GIS and remote sensing
  Scientists use tools like satellites and drones to gather information about the physical and chemical makeup of an ecosystem from a distance. For example, they can track the movement of animals like elephants and whales to learn about their habitats and protect them from dangers.
- Ecological modeling
  Scientists use math and computer programs to make predictions about how an ecosystem will change in response to different factors. These models simulate the interactions and processes within an ecosystem. For example, they can predict how a warmer climate will affect the plants and animals in an ecosystem.
- Field measurements
  Scientists can also gather data by going out into the field and observing and counting the different species in an ecosystem or measuring things like the amount of nutrients in the soil.
- Laboratory analysis
  In the lab, scientists can examine samples like soil, water, and plant tissue to understand how things like pollution are impacting the ecosystem.
- Citizen science
  Scientists also involve the public in data collection and monitoring efforts through programs called citizen science. This helps gather a lot of data and increase understanding of the ecosystem.

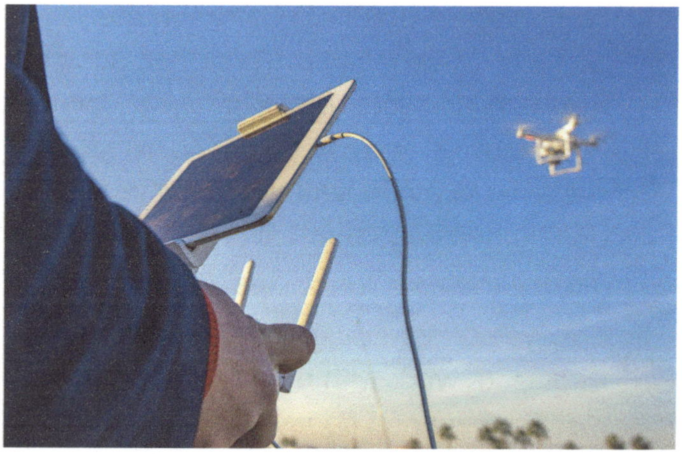

Engineer using drone to inspect land area

These tools help scientists understand how ecosystems work, find potential problems, and take steps to fix them. The goal is to keep the ecosystem healthy and strong.

Case Study: Chesapeake Bay

Chesapeake Bay

Chesapeake Bay is an estuary that borders Maryland and Virginia. It is the largest estuary in the United States and home to many different plant and animal species. However, the bay has faced a number of challenges in recent years, including pollution, habitat loss, and the impacts of climate change. These challenges have led to a decline in populations—such as the bay's two most important organisms, blue crab and oysters—negatively impacting the livelihood of those who depend on them.

The key factors that caused the bay ecosystem to decline were the following:

- runoff
- too much sediment in the water
- overfishing
- storms

A team of engineers has worked to restore and protect the bay's ecosystem. One issue they addressed was nutrient pollution in the bay. Excess nutrients had led to an increase in algal blooms that depleted the oxygen in the water. A study showed that most of the organisms in Chesapeake Bay did best in water that had oxygen levels of 5.0 milligrams per liter (mg/L) **(Fig. 8.1)**.

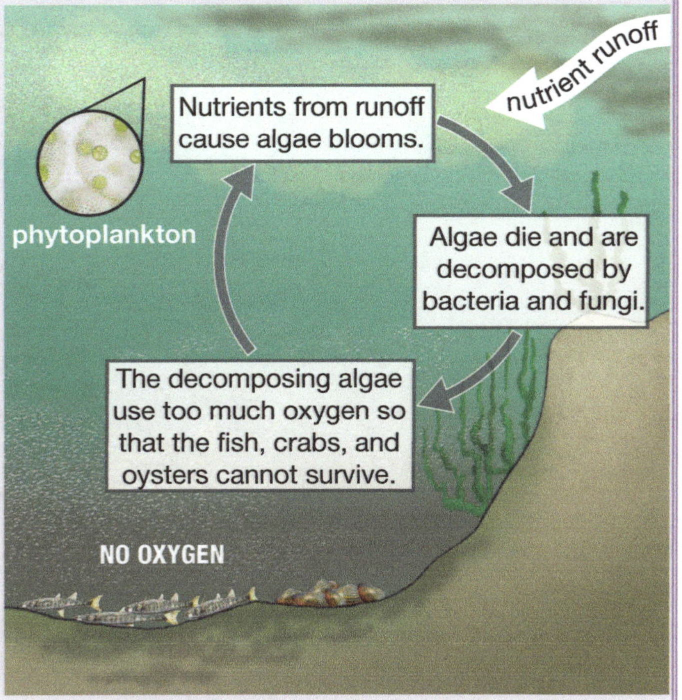

Over time, engineers helped to protect and restore the Chesapeake Bay by:

- redesigning wastewater treatment plants
- working with farmers to reduce pollution from wastewater and fertilizers
- stopping sediment and nutrients from running off the fields
- placing more oysters in the bay to filter pollutants
- planting native grasses and trees for wildlife
- monitoring the bay's ecosystem health with sensors and technology

These efforts have helped improve the water quality and increase the populations of species in the Chesapeake Bay. The team project serves as an example for protecting and restoring other ecosystems.

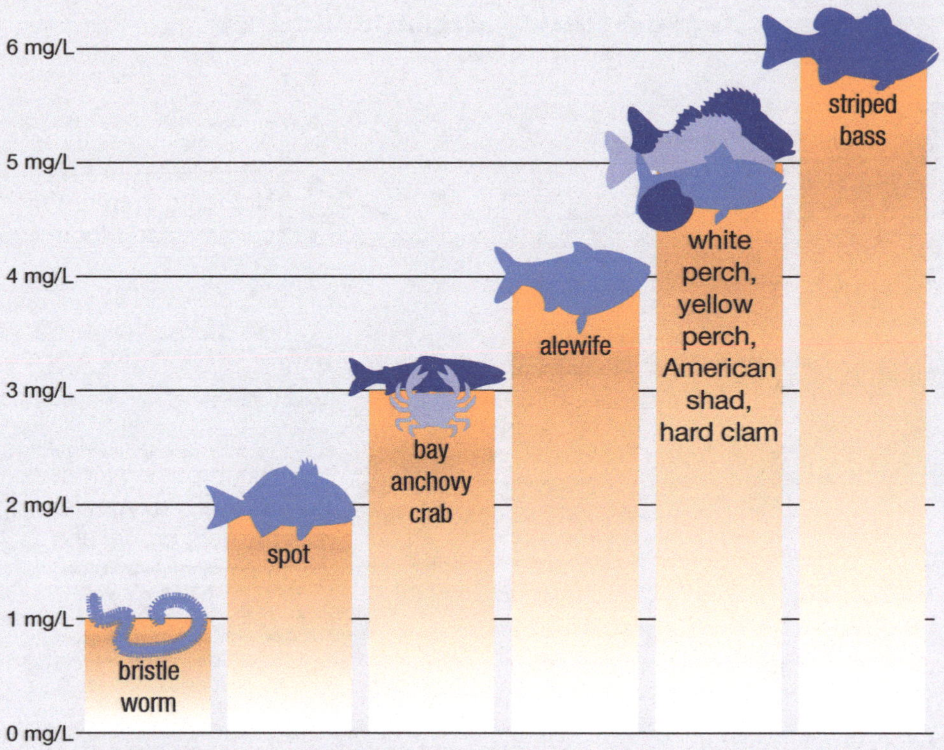

Fig. 8.1 Dissolved oxygen levels and marine life in Chesapeake Bay

REVIEW

Match each tool or method with its most likely use.

8.1) _____ citizen science programs

8.2) _____ ecological models

8.3) _____ field measurements

8.4) _____ GIS and remote sensing

8.5) _____ laboratory analysis

A. used to gather data on the physical and chemical characteristics of an ecosystem from a distance

B. use math and computer programs to predict how ecosystems change in response to different factors

C. collected on the ground through observations and measurements in the field

D. examines the physical and chemical characteristics of different samples

E. involve the general public in data collection and monitoring efforts

Choose the best answers.

8.6) _____ What is an example of an engineering project that addresses climate change?
A. developing irrigation systems that can survive droughts
B. building dams and levees to control the flow of water in an ecosystem
C. designing and building new roads
D. developing precision irrigation systems

8.7) _____ Which type of engineering project involves designing and building dams, reservoirs, and treatment plants?
A. ecosystem restoration
B. climate change adaptation
C. water management
D. sustainable agriculture

8.8) _____ What is the main goal of precision irrigation systems?
A. to control the flow of water in an ecosystem
B. to apply the exact amount of water, fertilizers, and other nutrients to crops
C. to protect and preserve ecosystems that are at risk
D. to monitor the health of ecosystems

8.9) _____ What is the main function of environmental monitoring tools and techniques?
A. to track water quality, air quality, or wildlife populations
B. to use fewer resources and have less of an impact on the environment
C. to protect and preserve ecosystems
D. to build infrastructure such as dams of levees

8.10) _____ What are some examples of remote sensing technologies that scientists use to study ecosystems?
A. satellites and drones
B. microscopes and telescopes
C. radios and televisions
D. cars and buses

8.11) _____ How can ecological models be used to understand an ecosystem?
A. by simulating the interactions and processes within an ecosystem
B. by observing and counting the number of different species in an ecosystem
C. by measuring the levels of nutrients in the soil
D. by examining the physical and chemical characteristics of different samples

8.12) _____ Which type of data collection is done by observing and counting the number of different species in an ecosystem or measuring the levels of nutrients in the soil?

A. GIS and remote sensing
B. ecological modeling
C. field measurements
D. laboratory analysis

8.13) _____ What are the two most important organisms in Chesapeake Bay?

A. turtles and snakes
B. fish and birds
C. algae and plants
D. blue crabs and oysters

8.14) _____ What were some of the factors that caused the Chesapeake Bay ecosystem to decline?

A. natural disasters, lack of sunlight, and overpopulation
B. runoff, too many sediments in the water, overfishing, and storms
C. lack of water, pollution, and human activities

8.15) _____ How have the efforts of the engineering team had an impact on the health and resilience of the Chesapeake Bay's ecosystems?

A. The water quality has improved and the populations of many species have begun to recover.
B. The water quality has decreased and the populations of many species have disappeared.
C. The water quality has not changed and the populations of many species have not been affected.

8.16) _____ What are some of the technologies and practices that engineers used to restore the Chesapeake Bay?

A. increasing the fishing quotas
B. building dams, levees, and roads
C. reducing nutrient pollution, controlling sediment, and increasing oyster populations

CHECK ☐ CORRECT ☐ RECHECK ☐

ACTIVITY: Evaluating Engineering Projects

In this Activity, you will evaluate different engineering projects that address ecosystem changes.

Instructions:

1 – Research one engineering project that addresses an ecosystem change. Examples of ecosystem changes include climate change, habitat loss, pollution, and overfishing.

2 – Create a report that includes the following information about the engineering project you have chosen:

- a brief summary of the problem the project is trying to solve
- a description of the engineering project and its goals
- an evaluation of how well the project is solving the problem. Consider factors such as the project's impact on the ecosystem, its cost-effectiveness (Are the benefits worth the cost?), and its ability to be replicated or implemented in other places or on a larger scale
- a discussion of any challenges or limitations of the project
- suggestions for how the project could be improved or modified in the future

3 – Include at least three sources in your report, one of which should be a primary source such as a scientific journal article or a government report.

4 – Submit your report to the teacher for grading.

RUBRIC					
Criteria	**Exemplary 4**	**Accomplished 3**	**Developing 2**	**Beginning 1**	**Score**
Criteria Organization	The report is well organized and easy to follow and includes all the necessary information.	The report is mostly well organized and easy to follow but may be missing some information.	The report is somewhat difficult to follow and may be missing some important information.	The report is poorly organized and difficult to follow.	_____
Content	The report includes all the required information and is thorough and detailed.	The report includes most of the required information but may be missing some details.	The report includes some of the required information but is lacking in detail.	The report is missing most of the required information.	_____
Credible Sources	The report includes at least 3 sources, with one being a primary source, and all sources are relevant and credible.	The report includes at least 3 sources, with one being a primary source but one of the sources may not be relevant or credible.	The report includes fewer than 3 sources, with one being a primary source, or not all sources are relevant or credible.	The report does not include a primary source or any relevant or credible sources.	_____
Evaluation of Project	The evaluation of the project's effectiveness is well-supported by evidence and analysis.	The evaluation of the project's effectiveness is supported by some evidence and analysis but may be lacking in detail.	The evaluation of the project's effectiveness is supported by little evidence and analysis.	The evaluation of the project's effectiveness is not supported by evidence and analysis.	_____
If not attempted, score = 0.					Total _____/16

TEACHER CHECK ☐

CHARACTER SPOTLIGHT

PAY ATTENTION!

Sometimes, things that are important escape our notice. A person is discouraged, and we miss the signs. An opportunity presents itself, and we lose out. We read a textbook and still cannot pass the test. Why? Simply this: we are not paying attention. One of the keys to success in life is to learn to be observant. This means paying attention to the details and trying to understand what it all means. If we can learn to do that, we can learn to do anything.

9. CAREER SPOTLIGHT: ENVIRONMENTAL ENGINEERING

Environmental science and engineering jobs involve the study of the natural environment and the development of technologies and ways to protect and preserve it. **Table 9.1** shows some of these jobs and their duties.

TABLE 9.1 JOB OPTIONS

Occupation	Job Duties
Environmental scientist	Study how the natural environment affects and is affected by human activities
Environmental engineer	Design and build systems to clean up pollution and protect natural resources
Chemical engineers	Use chemistry and engineering to solve problems in manufacturing and other industries
Civil engineers	Design and build roads, bridges, and buildings
Conservation scientists and foresters	Work to protect and manage natural resources like forests and wildlife
Hydrologists	Study water and how it moves through the earth's systems
Agricultural engineers	Apply engineering principles to solve problems in agriculture
Marine engineers	Design and build ships, boats, and other marine vessels
Waste management engineers	Develop ways to reduce, reuse, and recycle waste

These jobs are important for ensuring the health of the earth's ecosystems, as well as safeguarding human health and safety. Environmental scientists and engineers work in a variety of settings and are involved in a wide range of activities. In this Lesson, we will explore the role of an environmental engineer and why it is important to help to protect and improve the environment.

WHAT IS AN ENVIRONMENTAL ENGINEER?

Environmental engineers work to protect and improve the environment. They use science and technology to find ways to reduce pollution, protect natural resources, and make sure that people have clean air and water. Some specific things that an environmental engineer might do include the following:

- building systems to clean pollution, such as water treatment plants
- helping companies and factories reduce the amount of waste they produce and the pollution they create
- working with government agencies to make sure that laws and rules to protect the environment are being followed

Engineering team surveying forest

- studying how climate change affects the environment and making plans to help communities prepare for it
- designing systems to manage and protect surface water resources, such as dams and reservoirs
- working with other engineers and scientists to find new technologies and ways to protect the environment

Environmental engineers typically have full-time jobs. They work many hours managing projects and making sure that tight deadlines are met. They may work in a variety of settings, such as offices, laboratories, or in the field. Some environmental engineers work for government agencies, consulting firms, or private companies. Some environmental engineers work for research institutions and universities conducting research on environmental issues. They may also teach classes on environmental engineering and related topics.

Becoming An Environmental Engineer

To become an environmental engineer, one usually needs a bachelor's degree in environmental engineering or a related field, such as civil engineering or chemical engineering. Additionally, a strong background in science and math is important. Environmental engineers use these skills to design and analyze environmental systems.

After obtaining a bachelor's degree, some environmental engineers may choose to further their education by earning a graduate degree such as a master's or PhD. They may also choose to become licensed professional engineers, which requires passing a difficult exam.

Job Outlook

According to the U.S. Bureau of Labor Statistics (BLS), employment of environmental engineers is estimated to grow 4% from 2021 to 2031, about as fast as the average for all occupations. The BLS states that the demand for environmental engineers will increase as the population continues to grow and industrialization expands, leading to more concern for the environment. Additionally, tougher environmental regulations and the need for sustainable infrastructure will also drive job growth. The median annual salary for environmental engineers was $96,820 in May 2021.

First Female Environmental Engineer

Dr. Ellen Swallow Richards is considered the first female environmental engineer. She was born in 1842 and graduated from Vassar College in 1870, where she studied chemistry and geology. After graduation, she worked as a laboratory assistant at Massachusetts Institute of Technology (MIT), where she was the first woman to be admitted to the school's laboratory. She went on to study at the University of Michigan. Later returning to MIT, she earned a degree at MIT in sanitary chemistry in 1873, becoming the first woman to earn a degree at MIT and in that field. Dr. Richards was a pioneer in the field of environmental engineering and made significant contributions in the areas of water and air pollution control, sanitation, and public health. She is known as the "mother" of the field of sanitary engineering and was a founding member of the American Society of Sanitary Engineers. Some of her notable achievements include the following:

Ellen Swallow Richards pioneered the field of sanitary engineering.

- She was a pioneer in the field of sanitary chemistry, which is the study of the chemical and physical properties of water, air, and soil in relation to public health. She established the first sanitary chemistry laboratory in the United States.
- She was an early advocate for safe drinking water and developed methods for analyzing and measuring water quality. She also conducted research on water purification methods, such as filtration and chlorination.
- She was a strong supporter of environmental conservation and advocated for the protection of natural resources, such as forests and wetlands. She also conducted research on the impacts of pollution on human health and the environment.
- She was instrumental in standardizing the measurement of water, air, and soil quality as well as in developing methods for testing and analyzing these materials.
- She is considered the founder of the field of home economics, which studies the management of resources in the home and the impacts of these resources on human health and the environment.

Dr. Richards was a trailblazer in the field of environmental engineering. Her work laid the foundation for many of the practices and policies that are used today to protect public health and the environment.

THE LIFE OF
ELLEN H. RICHARDS

BY
CAROLINE L. HUNT

WHITCOMB & BARROWS
BOSTON, 1912

Book written about the life of Dr. Ellen Swallow Richards

REVIEW

Fill in the blanks using words from the box below.

chemical engineers
agricultural engineers
hydrologists
marine engineers
civil engineers
environmental scientists
environmental engineers
conservation scientists and foresters
waste management engineers

9.1) ________________________________ protect and manage forests and wildlife.

9.2) ________________________________ study water and how it moves through the earth's systems.

9.3) ________________________________ design and build roads, bridges, and buildings.

9.4) ________________________________ study how the natural environment affects and is affected by human activities.

9.5) ________________________________ design and build systems to clean up pollution and protect natural resources.

9.6) ________________________________ use engineering principles to solve problems in agriculture.

9.7) ________________________________ develop ways to reduce, reuse, and recycle waste.

9.8) ________________________________ design and build ships, boats, and other water vessels.

9.9) ________________________________ use chemistry and engineering to solve problems in manufacturing.

Choose the best answers.

9.10) _____ What is the primary goal of an environmental engineer?

A. to create new pollution sources
B. to protect and improve the environment
C. to make a profit
D. to design buildings

9.11) _____ What is an example of a task an environmental engineer might do?

A. break environmental laws
B. create a new video game
C. design a system to clean up pollution
D. create a new pollution source

9.12) _____ What is an example of a system an environmental engineer might build to clean up pollution in surface water?

A. a dam
B. a reservoir
C. a wastewater treatment plant
D. a solar panel

9.13) _____ What level of education is typically required to become an environmental engineer?

A. high school diploma
B. associate degree
C. bachelor's degree
D. master's degree

9.14) In which FIVE of the following work settings would an environmental engineer likely be found?

☐ government agency
☐ consulting firm
☐ manufacturing facility
☐ construction site
☐ research institution
☐ college or university

9.15) _____ Who is considered the founder of the field of home economics?

A. Dr. Ellen Swallow Richards
B. Dr. John Smith
C. Dr. Mehmet Oz
D. Dr. Michael Johnson

9.16) _____ What field of study did Dr. Ellen Swallow Richards found?

A. home economics
B. statistics
C. humanities
D. anthropology

9.17) Which THREE ways did Dr. Ellen Swallow Richards contribute to the field of environmental conservation?

☐ conducted research on water purification methods
☐ advocated for the protection of natural resources
☐ developed methods for testing and analyzing soil quality
☐ worked for a manufacturing facility

CHECK ☐	CORRECT ☐	RECHECK ☐

MY SCORE

QUIZ 3

TEACHER INITIAL

(Each question, 4 points)
Match the terms with the definitions.

3.01) _____ scale

3.02) _____ proportion

3.03) _____ quantity

A. the amount of a particular resource or species within an ecosystem

B. the size of an ecosystem

C. the ratio of different components within an ecosystem

Fill in the blanks using words from the box below.

GIS and remote sensing	ecological models	field measurements
laboratory analysis	citizen science programs	

3.04) ________________________________ is used to gather data on the physical and chemical characteristics of an ecosystem from a distance.

3.05) ________________________________ use math and computer programs to predict how ecosystems change in response to different factors.

3.06) ________________________________ are collected on the ground through observations and measurements in the field.

3.07) ________________________________ is used to examine the physical and chemical characteristics of different samples.

3.08) ________________________________ involve the general public in the collection of data and monitoring efforts.

Match the job duties with the occupations.

3.09) _____ protect and manage forests and wildlife

3.010) _____ study water and how it moves through the earth's systems

3.011) _____ design and build roads, bridges, and buildings

3.012) _____ study how the natural environment affects and is affected by human activities

3.013) _____ design and build systems to clean up pollution and protect natural resources

3.014) _____ use engineering principles to solve problems in agriculture

3.015) _____ develop ways to reduce, reuse, and recycle waste

3.016) _____ design and build ships, boats, and other water vessels

A. environmental engineers
B. hydrologists
C. environmental scientists
D. conservation scientists and foresters
E. civil engineers
F. marine engineers
G. waste management engineers
H. agricultural engineers

Choose the best answers.

3.017) _____ A decrease in the population size of a predator will lead to a(n) ___ in the population size of its prey.
A. decrease
B. increase

3.018) _____ If the proportion of a certain species of fish to other fish species was significantly decreased, what would happen to the food web of an ecosystem?
A. There would be no change to the food web.
B. The population of other species that rely on that fish for food would also decrease.
C. The population of other species that rely on that fish for food would increase.

3.019) _____ If the quantity of a key resource, such as food or water, were to decrease, how would the ecosystem be impacted?
A. The ecosystem becomes more diverse.
B. The ecosystem becomes less diverse.
C. The ecosystem becomes more stable.
D. There is no change in the ecosystem.

3.020) _____ What is the main function of environmental monitoring tools and techniques?

A. to track water quality, air quality, or wildlife populations
B. to use fewer resources and have less of an impact on the environment
C. to build infrastructure such as dams of levees
D. to protect and preserve ecosystems

3.021) _____ What are some examples of remote sensing technologies that scientists use to study ecosystems?

A. microscopes and telescopes
B. satellites and drones
C. stethoscopes and thermometers
D. radios and televisions

3.022) _____ Who is considered the founder of the field of home economics?

A. Dr. Mehrot Oz
B. Dr. John Smith
C. Dr. Ellen Swallow Richards
D. Dr. Michael Johnson

3.023) _____ What is the main job of an environmental engineer?

A. to cause pollution
B. to protect and improve the environment
C. to make money
D. to design buildings

3.024) Which FOUR are examples of tasks that an environmental engineer might do?

☐ develop plans to help communities adapt to climate change
☐ find new technologies and ways to protect the environment
☐ design systems to clean up pollution in cities, towns, and rural areas
☐ break laws and regulations that help protect the environment
☐ help businesses find ways to reduce the amount of waste and pollution they create

3.025) _____ ___ is one method of water purification that Dr. Ellen Swallow Richards researched.

A. Filtration
B. Wind power
C. Solar power
D. Nuclear power

CHECK ______ CORRECT ______ RECHECK ______

STOP and prepare for the Unit Practice Test.

- Review the Objectives and Vocabulary for each Section.
- Reread each Lesson and its corresponding questions.
- Relearn each Lesson that you still do not understand.
- Review the Quizzes.

MY SCORE

PRACTICE TEST

TEACHER INITIAL

(Each answer, 3 points)

Match the definitions with the terms.

1) _____ region of land where all the water is drained into the same body of water
2) _____ underground area where the rocks, gravel, sand, and clay are completely filled with water
3) _____ underground rock formation that is able to hold groundwater
4) _____ type of pollution that comes from a single source
5) _____ type of pollution that comes from contaminated runoff
6) _____ process of converting chemicals into an energy form usable by organisms
7) _____ microscopic organisms that make up the bottom of the food chain in aquatic ecosystems
8) _____ the area immediately surrounding a lake or river that contains trees, grasses, and other vegetation

A. aquifer
B. water table
C. watershed
D. point-source
E. nonpoint-source
F. riparian zone
G. plankton
H. chemosynthesis

Choose the best answers.

9) _____ ___ helps the climate to be more moderate due to currents bringing warm water to colder areas.
   A. The ocean
   B. A river
   C. An estuary
   D. A lake

10) _____ Which type of aquifer is more easily polluted?
   A. artesian wells
   B. restrained
   C. unconfined
   D. confined

11) _____ Could an oil spill affect the fish populations in the ocean?
   A. Yes, fish populations would likely increase.
   B. Yes, fish populations would likely decrease.
   C. No, fish populations would likely stay the same.

12) _____ An ocean ecosystem is polluted with methylmercury. What process would cause killer whales to have the most methylmercury in their systems?

A. sedimentation
B. coral bleaching
C. chemosynthesis
D. biomagnification

13) _____ Washing a car on the road and using fertilizers can cause ___ to enter the water.

A. pathogens
B. nutrients
C. plankton
D. oxygen

14) _____ Groundwater is stored in ___.

A. rivers
B. reservoirs
C. lakes
D. aquifers

15) Choose THREE things that ocean acidification can cause.

☐ water temperatures to increase
☐ coral to die
☐ shells to be deformed
☐ oxygen levels to increase

16) _____ *Trash from the streets enters the storm drain when it rains. The storm drain empties into a river.*

What type of pollution is this?

A. watershed pollution
B. point-source pollution
C. nonpoint-source pollution
D. aquifer pollution

17) _____ *Trees keep water cool by providing shade.*

If the trees were removed, what would happen as temperatures increase? What would happen to the oxygen levels?

A. Oxygen levels would decrease.
B. Oxygen levels would stay the same.
C. Oxygen levels would increase.

18) _____ *Nutria are an invasive species in many rivers and wetlands in the United States. Nutria reproduce quickly and eat all the vegetation, leaving very little for other species, such as the native muskrat. They also destroy riparian zones.*

What would happen to the muskrat populations if nutria were introduced into this ecosystem?

A. Muskrat population levels would increase.
B. Muskrat population levels would decrease.
C. Muskrat population levels would stay the same.

19) _____ *Mangrove forests are found in estuaries.*

Mangrove trees have adapted to filter out extra ___ from the brackish water.

A. salt
B. oxygen
C. nitrogen
D. acid

20) _____ *Runoff that enters the river has chemicals that can change the pH of the stream.*

The graph shows the pH levels an organism can live in. Which organism can best survive changes to the stream if the water becomes more acidic?

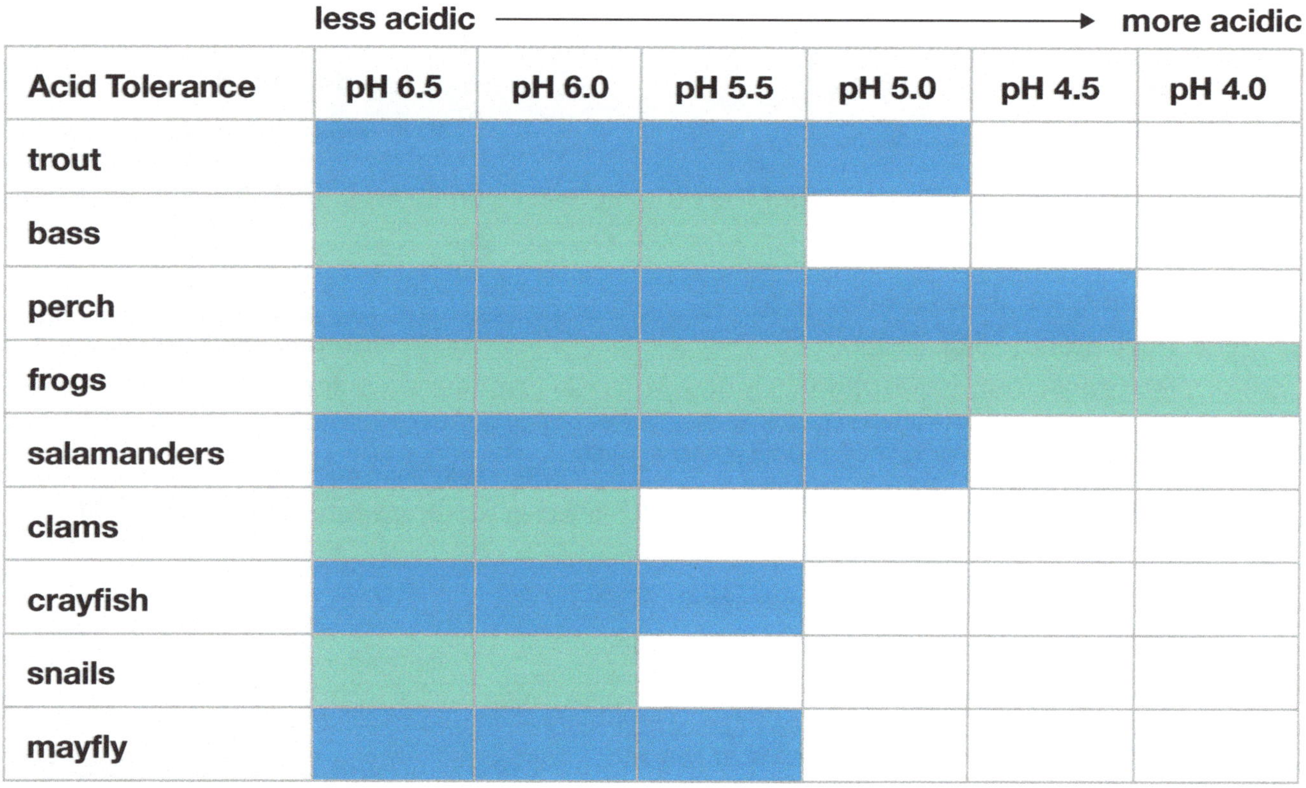

A. snails
B. trout
C. salamanders
D. frogs

Write the correct answers.

21) Order the algae bloom events using the numbers 1–4.

a. _____ Algae reproduce at a high rate.
b. _____ Excess nutrients enter the water.
c. _____ Level changes in water cause fish kills.
d. _____ Algae decompose and take up oxygen.

Choose the best answers.

22) _____ In ecology, the amount of a particular resource or species within an ecosystem refers to ___.
A. scale
B. quantity
C. proportion

23) _____ The size of an ecosystem refers to ___.
A. quantity
B. proportion
C. scale

24) _____ The ratio of different components within an ecosystem refers to ___.
A. proportion
B. quantity
C. scale

25) _____ A decrease in the population size of a predator will lead to ___ in the population size of its prey.
A. an increase
B. a decrease
C. no change

26) _____ If the proportion of a certain species of fish to other organisms was significantly decreased, the population of other species that rely on the fish for food ___.
A. would also decrease
B. would increase
C. would remain the same

27) _____ The ecosystem would ___ if the quantity of a key resource, such as food or water, were to ___.
A. become more diverse; increase
B. become more diverse, decrease
C. not change; increase
D. not change; decrease

28) _____ The main function of environmental monitoring tools and techniques is ___.
A. to track water quality, air quality, or wildlife populations
B. to protect and preserve ecosystems
C. to build infrastructure such as dams of levees
D. to use fewer resources and have less of an impact on the environment

Match each tool or method with its most likely use.

29) _____ laboratory analysis

30) _____ ecological models

31) _____ field measurements

32) _____ GIS and remote sensing

33) _____ citizen science programs

A. used to gather data on the physical and chemical characteristics of an ecosystem from a distance

B. examines the physical and chemical characteristics of different samples

C. uses math and computer programs to predict how ecosystems change in response to different factors

D. involves the general public in data collection and monitoring efforts

E. collected on the ground through observations and measurements in the field

CHECK ____ CORRECT ____ RECHECK ____

You must now prepare for the Unit Test.

- Review the Objectives and Vocabulary for each Section.
- Reread each Lesson and its corresponding questions.
- Review and study the Quizzes and Unit Practice Test.

When you are ready, turn in your Unit and request your Unit Test.

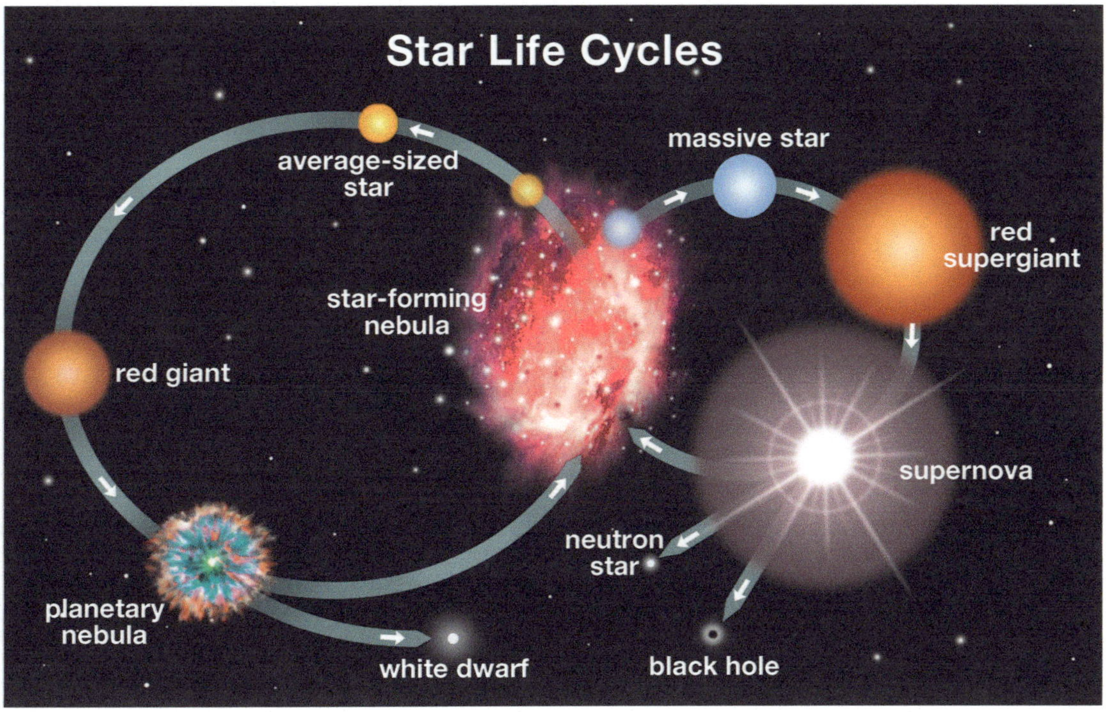

Parts of a Wave

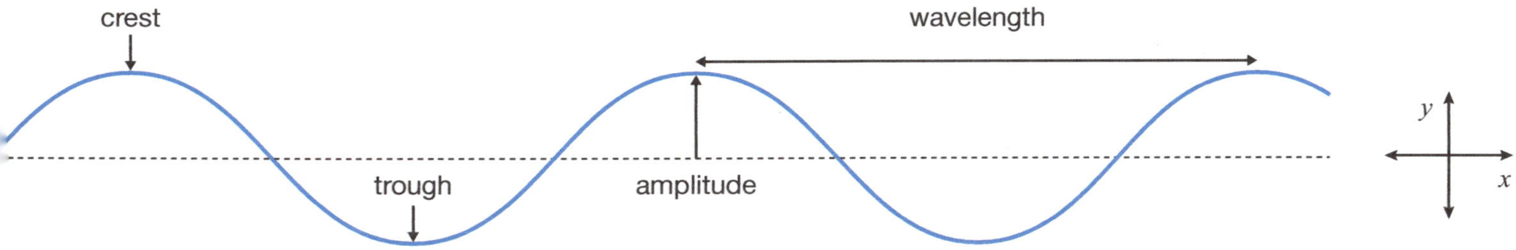

Natural Greenhouse Effect

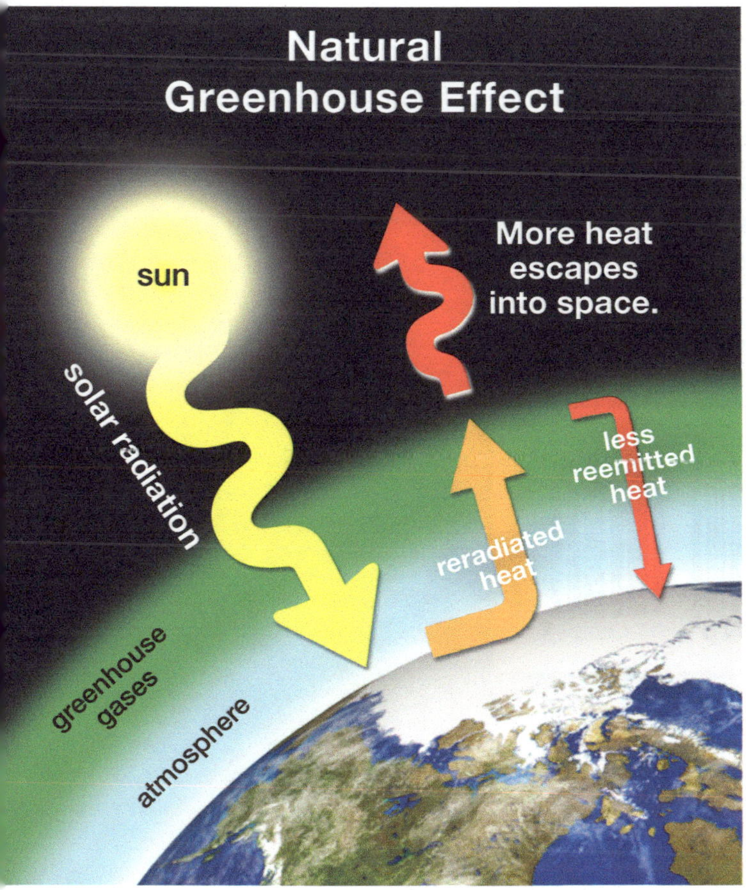

Human-Enhanced Greenhouse Effect

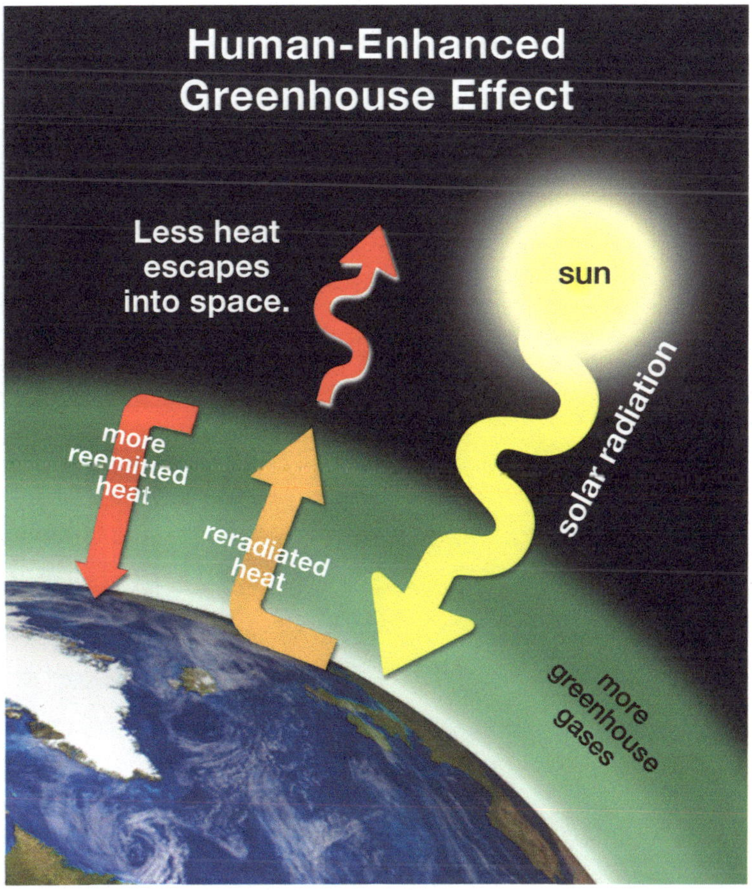